The Crossing of Heaven

Karl Gustafson

The Crossing of Heaven

Memoirs of a Mathematician

 Springer

Karl Gustafson
University of Colorado
Dept. of Mathematics
Boulder, CO 80309
USA
Karl.Gustafson@Colorado.EDU

Math.Subj.Classification (2010) : 00-XX, 97-03, 01AXX, 68-03

ISBN 978-3-642-22557-4   e-ISBN 978-3-642-22558-1
DOI 10.1007/978-3-642-22558-1
Springer Heidelberg Dordrecht London New York

Library of Congress Control Number: 2011942157

© Springer-Verlag Berlin Heidelberg 2012
This work is subject to copyright. All rights are reserved, whether the whole or part of the material is concerned, specifically the rights of translation, reprinting, reuse of illustrations, recitation, broadcasting, reproduction on microfilm or in any other way, and storage in data banks. Duplication of this publication or parts thereof is permitted only under the provisions of the German Copyright Law of September 9, 1965, in its current version, and permission for use must always be obtained from Springer. Violations are liable to prosecution under the German Copyright Law.
The use of general descriptive names, registered names, trademarks, etc. in this publication does not imply, even in the absence of a specific statement, that such names are exempt from the relevant protective laws and regulations and therefore free for general use.

Cover photograph of the author by Jillian Lloyd

Printed on acid-free paper

Springer is part of Springer Science+Business Media (www.springer.com)

"Who can know one's own destiny?

To the Fabric of Our Lives
And to its Richness and Beauty

Foreword

A few days before his arrival in Greece for the Web Science Master Course Lectures, Karl Gustafson asked me if I would write the Foreword to his autobiography. I agreed readily, of course, considering it a great honor and a personal pleasure. Moreover, I was excited by the prospect, for three particular reasons:
1. I have known and admired Karl for many years: We first met in 1984 when he visited Brussels, where I was working toward my Doctorate Thesis on Time Operator, Irreversibility and Complex Systems with Ilya Prigogine and Baydyanath Misra. With Karl Gustafson, the trinity of my thesis advisors was complete. Since then, I have been privileged to collaborate and discuss with him several conventional as well as unconventional issues, including Mathematics, Computation, Physics, Biology, Cognition, Knowledge, Economy, Philosophical Problems, and Social Dynamics. One of our resonant interactions during the ensuing 25 years was my urging him to write his autobiography, especially for the younger generations, not only because it would be an enjoyable story to read, but mainly because one can easily identify two remarkable qualities of Karl. The second and third reason for my eagerness to write this Foreword are precisely these qualities.
2. Karl Gustafson is a living synthesis of three persons: an Engineer, a Physicist and a Mathematician. During his university studies, presented in Chapter 3, "The Student in Poverty", and throughout his life challenges, Karl managed to operate simultaneously in the three different modes, and get the best out of each in any practical and theoretical matter he was engaged in. This is a rare practical demonstration of multidimensional spirit.
3. There was no island of stability in Karl's scientific journey. Rather, Karl successfully crossed among different disciplines and professional communities, including Pure and Applied Mathematics, Computational Mathematics, Theoretical Physics, Fluid Dynamics, Optical Computing, Neural Networks, Mathematical Finance, and Complex Systems, demonstrating the natural link between different scientific domains. Accordingly, Karl developed a practical realization of the necessary trade-off between width and depth, as expressed in his Generalization-Specialization Paradox, discussed in Chapter 11, "Mathematics".

Why did Karl not follow the widely "recommended" career path to restrict himself to one mainstream field and thereby profit with a socially "better" professional profile? Why did he jump from Engineering to Physics and Mathematics?

The answer is, because he had reached the limits of the procedural knowledge of "how to do things" and wanted to understand "why things and events emerge as such." In this process, Karl experienced the natural beauty of knowledge, which cannot be split to fit into our specific professional disciplines. Karl instead demonstrated the unity of nature, in contrast to our fragmented representation of reality. This is nice to say but hard to experience. Facing the dilemma of power and high social recognition versus the satisfaction of the appetite for knowledge, Karl chose the second. Is this not the quality of a hero?

During his Web Science Course, Karl presented his work on Intelligent Distributed Systems in four lectures, which captivated and thoroughly impressed the students. First, he viewed the Satellite Revolution presented in chapter "Computers and Espionage" as the first stage of the Web Science revolution. Then he showed how his Human Trajectory Analysis discussed in Chapter 13, "The Improbabilities" was a precursor of the Small World Networks and Google Algorithm. Modeling Humans versus Neural Nets and machine learning, Karl realized early on the need for Context and Simplification, anticipating present developments of the Semantic Web and Ontologies. The last lecture involved Finance Networks, presented in Chapter 12, "High Finance", with several remarks on the present financial crisis and the emerging Web Economics.

Here Karl has written a tale that will charm and entertain the reader, while inspiring the mind and soul. The innovative spirit, fresh ideas, and thought power of Karl Gustafson indicate already that there will be a further "Crossing of Heaven" in a few years.

Ioannis Antoniou
Professor of Mathematics
Aristotle University of Thessaloniki
Thessaloniki, Greece

Preface

I am not a famous mathematician, nor a Nobel Prize winner. I am well known in the sphere of world-class mathematicians – both at home in the United States and more-so internationally, where I have worked at the highest levels throughout my career. But that is not why I chose to write this book, this "story of my life," if you will. No, it has little to do with my being famous, or infamous. The seeds of my desire to write an autobiography were first planted by friends and colleagues, who long urged me: "Karl, you should write a book about your life! You have so many tales to tell." And I do have stories to tell that have never been told. Experiences and events that I wish to share, so that they are not lost to history. Then five years ago, a much respected junior colleague in Greece (with whom I share special bonds of being long ago his Ph.D. advisor, and then for many years as scientific collaborator) put the pressure on me. He wrote: "Karl, you express a talent for fresh/ historical/ philosophical writing which allows you to insert much easier messages not only to scientists but also to intelligentsia in general. Do it now! Write the book of your life! Everything you have done so far is just preparation. Sorry if I was too spontaneous."

I was of course much moved by this honoring of my life and affirmation of our long friendship. I vowed to get to the task. But the inertias of a busy professional life kept me fully occupied in those directions.

The tipping point came 2 years ago when at a class reunion I met an old college classmate who had risen to the top levels in the intelligence agencies. In confidence I told him that I too had been thrust into intelligence work of great national importance early in my career. Those special and unusual circumstances led, in what I regard as an accident of fate, to my writing the software for the world's first spy satellite. He replied: "Karl, I have seen that satellite. The program was recently declassified. Do you want me to send you a picture of it?"

Indeed, it seems that I have witnessed – and been an integral part of – the technological revolution that has transpired during the last 50 years. *En plus,* my mathematical and scientific interests have been wide and varied, resulting in many exciting plenary speaking invitations through which I have traveled and seen much of the world. I must add to those two components, those of active participant in an astounding scientific revolution, and of world traveler, a third and more personal one: I am hardly your rich or Ivy League type, or as I sometimes too irreverently refer to them, the "effete-elite" scientific type. My life has had hard knocks and brushes with death and as a consequence I most closely identify with the common man.

I measure this incredible scientific and technological revolution which has forever transformed our lives as starting about 50 years ago on October 4, 1957, when the Soviet Union took humanity into the Space Age with their launch of Sputnik, the first artificial Earth Satellite. A few days after, I and a few other physics students ran outside from our Physics honor society banquet on a crystal clear night high in the Colorado mountains to watch Sputnik trace its orbit across the heavens above us. I can remember that night, and the thrill and the drama, still.

The United States went into panic-mode and in classic American fashion, united to fight back, to compete. I was a direct beneficiary. Mathematics, especially applied mathematics, was deemed of paramount importance. Even 30 years later when I lectured at Moscow State University, there were many military officers of high rank in the audience, testament to the very close links between the academics there and Soviet military science. Sputnik and the critical scientific and political challenges it brought dramatically to the fore caused money to pour into mathematics and mathematics education in the United States. As a result I was one of a contingent of 12 top engineering students who were selected and paid to become Applied Mathematics undergraduate Engineering Problems Instructors. We taught all engineering first-year students the mathematics of slide rule and mechanical calculators. We had our own individual classes (a lot of them...) and that early initiation into university-level teaching not only built my confidence but also soon converted me from an engineering physicist to an applied mathematician. Some years later I would go all the way to a pure mathematician.

There is a substantial, nontrivial, and not widely understood gap between the training needed to become a pure mathematician and that of just being an engineer or physicist. This gap cannot be fully appreciated by applied scientists unless and until they successfully complete the committed step of taking many course-years of algebra, topology, measure theory, geometry, and real analysis, among others.

Sputnik and the ensuing satellite and more general Space Race not only changed my life, but it is fair to say that for a time, it captured my life. Happily ensconced as a full-time Instructor of Applied Mathematics at the University of Colorado, and starting to enjoy skiing and the good life, a strange sequence of events led to my being drafted into the U.S. Army Reserves so that I would be forcibly assigned as a young computing expert to a top secret U.S. Navy military espionage project for four years. It was within that environment that in a furious two-week effort I wrote the software for the world's first spy satellite, a project so secret that I was not even officially cleared to know of it.

As a consequence of this fantastic and favorable situation within which I found myself, I was permitted at the same time to pursue a Ph.D. at the nearby University of Maryland. I chose Mathematics over Physics. Thereafter I embarked on two postdoctorals in Europe. Eventually I returned to Colorado as a Professor of Mathematics. The rest, as they say, is history.

That history was, however, far from mundane. During my lectures in Colombia I was under constant threat of being kidnapped. In Russia a young former Soviet

Union general and I almost fell on the floor when we broke out laughing together about our open discussion of stealth technology. In India I was thrown together with a Fields prize winner as fellow refugees, as religious violence swept that entire subcontinent. In Tibet I traveled with an American ex-CIA agent who exactly five years earlier had been in Tiananmen Square the day it happened. My invited keynote lectures in Iran were thwarted two times by visa cancellations contrived by the Conservatives but at the last minute the Liberals prevailed, and upon discovering my birth date, they delightedly insisted on having not one but two fabulous birthday dinners for me. Twice I have found myself a single parent with sole responsibility for the care of young children. I stopped rock climbing at age 50 after a dangerous pendulum off-route on a high peak exactly three weeks after my girlfriend had died on a climbing trip to Peru.

Still, two years ago, I could not begin this autobiography. I was, in the vernacular, just plain stuck. One of my favorite early reads was *Zen and the Art of Motorcycle Maintenance* by Robert Pirsig. Being stuck is part of the process. To deal with it, you must care, and keep on caring, as you think on solutions. What was wrong? Here I was, the most prolific mathematician in the history of Colorado mathematics, with 270 published papers, 5 to 15 books depending on how you count, over 100 plenary addresses at international conferences in 35 countries – and I could not even begin my own autobiography.

When one is stuck, one should consult friends. Also, it is okay to lower that pride-shield and admit to needing a little help. Out of a casual lunch came an offer of more than a little help. The telling of my life story would not have happened without the editorial assistance of Jillian Lloyd, who got me started and held my hand throughout the process (and at times, the ordeal) of writing. Any man would be inspired by the attention of such a beautiful and intelligent young woman.

Once when joking with her as she insisted on raising the quality and depth of some passage in my exposition, I joshed, "Look, *cherie*, I'm just a poor scrap of a fun-loving western boy, with some serious sides and unique insights, who found himself in some very unusual situations." But she would not let me get by with such excuses. Gradually it became clear: I am not introspective. In particular, like many males, I do not want to deal with my emotions, I do not want to ruminate upon them. Rather, I am extrospective: I am interested in the world. These tendencies are further enhanced by my scientific training. So whatever introspective thoughts you will find in this book were mostly coaxed out of me. At times to get them out was tough going...but they are part of me and therefore an essential part of the book.

For me, it was much more enjoyable spinning the tales of the adventures of my life, the action, the close calls with death, the women, the science, the mathematics, and how my life interacted, many times unexpectedly, with important world events and Nobel prize winners and other famous and sometimes strange personalities. Events and accomplishments hold their own interest, but it is in the flavor of human interactions that our lives find their true meaning. You will find many in this book. – Karl Gustafson.

Contents

1. **The Child in Iowa** 1
   ...a cold, bright, sunny day...............................
2. **The Boy in Boulder** 11
   ...and a lot of rock climbing..............................
3. **The Student in Poverty** 19
   ...survival and success....................................
4. **Computers and Espionage** 27
   ...and the world's first spy satellite.....................
5. **First Publication** 37
   ...and a very strange letter...............................
6. **Into Academia** 49
   ...despite the politics....................................
7. **The World Opens** 57
   ... a community of scholars ....
8. **Personas and Personalities** 103
   ...and a penetrating theorem...............................
9. **Wives, Lovers, Friends** 113
   ...and a great loss..
10. **Close Calls** 125
    ...in particular, the lightning strike....................
11. **Mathematics** 133
    ...and some science.......................................
12. **High Finance** 145
    ...and dangers of overquantization........................

| 13. | **The Improbabilities** | 155 |

...and the imponderables...

| 14. | **Realities** | 167 |

...and time is running out...

| 15. | **The Crossing of Heaven** | 175 |

1. The Child in Iowa

...a cold, bright, sunny day...

I was born in the middle of the economic Great Depression on May 7, 1935, in the small town of Manchester in the slightly rolling hills of northeastern Iowa. My first vivid memory is that of being outside pulling my wagon along the sidewalk on a cold, bright, sunny winter day. I was 3, alone, and very happy. My mind was clear, and I was struck by the beauty of the world.

My entry into this world had been in a small house on Butler Street. The image of that cold, bright, sunny day, there on Butler Street, comes to me often. I see a small yellow house with a black coal-bin on the side, surrounded by not very deep snow. The sidewalk has a few barren oak or elm trees at intervals along it. I see only the trunks, quite dark against the snow. I am warmly dressed in my black sheepskin coat and helmet hat with ear flaps tied under my chin. The coat had those big buttons and thong fasteners, rather than button holes, and probably I had those old-fashioned galoshes over my shoes. Certainly that was the first time I had been allowed to venture out by myself. As I felt the crisp wind frosting my nose, and as I saw the black of the trees, the yellow of the house, and the brilliant blue of the sky overhead, I first became consciously aware of the world. Probably that was as happy as I have ever been.

Now it is another cold, bright, sunny day many years later and I am in Boulder, Colorado, in the foothills of the Rocky Mountains. The world is in the midst of another economic depression, and I am happy. Much has happened, but my mind and spirit are still refreshed as I take a snowy path in the green of the pine trees under an incredibly deep blue sky. There is a sense of harmony with the universe.

My thoughts go back to my childhood days in that simple life of Manchester, a small town of about 3,000 folk on the Maquoketa River. Iowa was Heartland America and like many small towns in Iowa at that time, its main function was to host the local small businesses that supplied the surrounding farming community. The summer days and nights were hot and humid. Winter days and nights were cold. In the fall we all raked the leaves into the curbs on the street and then burned them, roasting marshmallows on slender sticks. In spring we waited for the thaw in which the river ice broke free and piled up into great dams before the pressure behind broke those free too.

I was the second child and second son of Edwin Gustafson and Jeanette Anderson Gustafson – both of whom were born early in the 20th century to Scandinavian immigrants in America. My paternal grandfather, Carl August Gustafson, was born in Rya, Sweden, a tiny place about 40 km south of Laholm on the West Coast. At a young age, he was apprenticed out as a tanner. There, he was beaten badly, and ran away to Germany at age 16. He found a job sweeping floors at an eating establishment, where he learned to cook. He soon emigrated to America, arriving at Ellis Island before the Statue of Liberty had been installed. He then walked from New York to Virginia to look for work on the railroad. Having no money, he survived by eating sweet potatoes he found in the fields. In spite of speaking no English, he soon obtained work on the railroad. One night, the cook for the railroad got so drunk that Carl replaced him. Before long, he was contracting all the food purchases himself, and making a personal profit of $27,000 for three months work.

In Chicago, my grandfather met my grandmother Sofie Paulson (née Anna Sofia Palsdotter), who had been born in Vaxtorp, Sweden, a town about 20 km south of Laholm. She had emigrated to America at age 16 for a job as a domestic helper with a Swedish family in Chicago. They met at the urging of their families in Sweden, who knew each other. By around 1890, they married. As Carl continued to work as a cook and food supplier to the railroad crews heading south and west from Chicago, he amassed a small fortune. He and Sofie, who by then had two daughters, decided to "retire" back to Laholm, Sweden, in 1897. There, a third daughter was born, and Carl developed a drinking problem. By 1900, with most of their money gone, Sofie convinced Carl to return to America. At first, he worked in coal mines and brickyards, and later returned to his specialty as cook and food supplier to the railroads pushing West. They finally settled down in the cowboy town of Plaza, North Dakota, where Carl built his own hotel in 1907. He and Sofie went on to have six children, four daughters and two sons, including the youngest, my father Edwin, born in 1910 in Plaza.

Their hotel burned to the ground in 1935 and my grandfather Carl decided to retire again, this time to North Hollywood, California, where he bought a city block. But he lived only two years more, and to my knowledge, we never laid eyes on each other.

Back in Plaza, North Dakota, my father Edwin Gustafson had been senior class president and a hot basketball player. Afterward, he had been sent to Carleton College in Northfield, Minnesota. He majored in economics and became a top intercollegiate debater. He also met my mother, Jeanette Anderson, a music major preparing for a career as a concert pianist. But soon after, she became pregnant – and they quickly married. He graduated, she did not; and he faced the daunting prospect of supporting a family in one of the worst economic times in the history of this country. Eventually, in 1933, with a $3,000 loan from his father, he bought the Gambles Hardware store in Manchester.

My brother Dick, born three years before me in 1932, has often helped me remember some of the circumstances of our childhood. He always believed that our parents had not wanted him. I have no idea at all if I was wanted. Nor have I ever worried about it. But it always bothered my brother Dick.

As for my childhood in Manchester, Iowa, I have only happy memories. My father worked hard long hours at his Gambles Store and some other related enterprises. He was prosperous, and my mother kept house and raised us. I still remember Sunday afternoons when the four of us would go down to the Castle Café and have a nice dinner served in that special private back room of the restaurant. This signaled his success and our upper middle-class status in that small town.

I do not remember a particularly cheerful family life at home. But remember, this was a WASP family, and displaying emotion was not encouraged. I don't recall much about the first five years of my life. We moved from Butler Street to a better house on Fayette Street. That was Depression-era 1935–1940. I remember the "tramps" who would come down the alleys behind the houses and knock on the back doors, asking for food. They were all unemployed men who would ride the railroad into town, try to get something to eat, and then move on. You never saw them on the main streets, only in the alleys. They were invariably very polite, and no one was afraid of them. My mother always gave them food.

My parents, and probably most of the businessmen in town, were strong Republicans. They hated Roosevelt and there were mean political jokes about him. Here is one I still remember: "I will move the Atlantic to the Pacific, and the Pacific to the Atlantic, and fill the Grand Canyon with beer. Of course, there will be a slight tax." The word "beer" was pronounced with that distinctive F.D.R. accent, like "beeehhhrrr." Has nothing changed?

I remember many things from the war years 1940–1945. I and my friends read all of the comic books of the era. There were the "Huns" (Germans) and "Nips" (Japanese), awful enemies always doing terrible things to our wonderful American soldiers, and bayoneting babies in China. I learned my first German swear-word – "Schweinhund!" – as the German soldier bayoneted the American soldier. We kids would go to the Isis Theater on Saturday mornings to watch Tom Mix and Hopalong Cassidy cowboy movies. There were always short war-documentary news programs before the main feature. The country was awash in propaganda. My brother Dick and I created little war games by cutting out paper soldiers from newspapers and comic books. We would then array them on the fine oak floor in the bedroom we shared, and one side would attack the other. Killing was done by pounding on the enemy soldier with your little automatic pencil with no lead in it. This would punch little "bullet holes" in the enemy paper soldier. It would also leave the fine oak flooring with thousands of little holes punched into it. Our parents were furious when they discovered the result of our, "Oh, they are nicely playing in their room."

My best friend was Johnny Whisler, who lived down the alley, a few houses away. Other friends were Russell Widner, and later Billy Post, Billy McCormick and Gordon Smith. Russell claimed my mother had to wait on May 6 for Dr. Jones to deliver him first; he was slow coming out, and that is why I was not born until May 7. There is an obvious medical fallacy to that story, but Russell went on to become a medical doctor. His father had only been a pharmacist.

The Widner drug store was still there when I visited in 2003. The Whisler men's clothing store had been taken over by another small retail business. Amazingly, my father's hardware store was still there. I visited with the owner, who was not young. I was surprised to see how small the store was; it seemed large in my childhood. I went down the stairs to the dark little basement where my father hired John Whisler and I to sit putting nuts on bolts for something like a penny a box. Probably that was my first "job." I also had a paper route. Saturday mornings were unpleasant as I rode my bike from house to house, trying to collect from evasive subscribers.

Schools in Iowa in that époque were considered the nation's best. Education was highly valued by the sons and daughters of poor immigrants, who were mostly of Scandinavian and German descent. The school classes were very small. There were only about 20 of us at each grade level. Our lives as children centered around two venues: the school, and the river. I loved the Maquoketa River and would take long hikes by myself after school and on weekends, exploring its sandbars and adjoining forests.

It was in the sixth grade that I felt my first attraction to a member of the opposite sex. Of course there had been some girl with whom I would always wrestle during recess. But I regarded her as a combatant, not as a girl. Then there were always the "spin the milk bottle" evenings with the neighborhood girls our age. If the bottle pointed at you, you had to kiss someone of the opposite sex. I always chose to kiss a cute little girl named Shirley. But the real event was my noticing Jackie Maulson, her legs, and her pretty Irish good looks. Eventually she invited me one day after school over to her house. Looking back, I was a total innocent. No one was home, and she sat me down on the sofa and gave me a scrapbook to look at. It may seem strange, but I can still feel her standing warmly next to me, then sitting down very close to me. I am not sure what would have happened next, but rather soon my father showed up and politely but firmly got me out of there. But my intuition and preferences had been right on target. Jackie was a beauty and went on to become Senior Prom Queen and eventually an exotic dancer.

Even as a child, I had strong intellectual inclinations. When my parents bought the obligatory encyclopedia set – it was called Book of Knowledge – I would lie on the living room floor for hours reading it. I still can see the page with the airplane flying at 200 mph toward the planets and stars, and how many years it would take to reach each one. At night after dinner I would bicycle down to the small Carnegie Library. There, alone with the gentle old librarian, I would read for hours. I liked an author named Howard Pease. He wrote adventure stories with titles like "The Black Tanker" or "Secret Cargo." The librarian would greet this little bookworm as he parked his bike and entered grinning: I can still remember her saying, "I have a new book for you, would you like to read it?" I would eagerly assent.

I remember very clearly sixth grade, where Mrs. Hansen had us learn to diagram sentence structure. It was easy and natural for me. When no one else could dissect a sentence, she would call me to the blackboard and usually I got it broken apart very quickly and correctly. Was this a precursor to my later-evident mathematical abilities?

* * *

I am three-quarters Swedish, and one-quarter Dutch in ancestry. My mother was half-Swedish and half-Dutch. During my Scandinavian mathematics lectures tour in 2002, I decided to take some extra days to hopefully trace my heritage in Sweden. In Laholm, the local librarian and Lutheran church records rewarded me with a documented history of the Gustafson and Paulsson lineage back to 1629. The librarian even found that she and I have a common ancestor named Sevid Olsson nine generations back! But what about my mother's side? There the story has a curious twist.

Before my lectures in Uppsala in 2002, with great generosity the Swedish mathematicians Owe Axelsson and Maya Neytcheva met me at the Stockholm airport and we set out to spend two or three days in the Askersund region at the north end of Lake Vättern. Our goal was to research the story of Andrew P. Anderson, my mother's Swedish grandfather. He and his wife, originally Clara Johansdotter, had emigrated to Wilmington, Delaware in 1872, where they established a Baptist church. A Baptist from Sweden? Strangely enough, that seemingly minor schism in my Swedish background would later cause considerate strife in my and my brother Dick's childhood.

The story, with some documentation to support it, was that Andrew was the illegitimate baby of the maid and the rich son of the wealthy family for whom she worked. That family owned an ore-processing factory on a small lake near Asker – and lived in a mansion house above it. We actually found the lake, and house where my great-grandfather was probably conceived. The place is called Bystad, and although privately owned, I would call it a Swedish national treasure. We walked about 500 meters up a country lane to find it. Lo and behold, there you find a perfectly preserved, approximately 200-year-old mansion, a separate chapel building, and a short distance away, a preserved 400-year-old Swedish residence.

Andrew's mother died only six months after his birth, and he was raised as a foster child by several families. Presumably this is how he picked up the name Anderson. At age 12 he was sent out to work in the fields, as was the custom in those times.

Nearby, we even found the Baptist church to which my great-grandfather had belonged. It is still in service and we heard the choir singing as we quietly visited the place on a Sunday morning. How could my great-grandfather know that his being a Baptist would impact his great-grandson on a separate continent some hundred years later? But it did. In Manchester, Iowa, my Lutheran father and Baptist mother fought, sometimes bitterly, over whether my brother Dick and I should go to the Lutheran or Baptist Sunday School. Neither gave in, so finally we were sent to the Congregational Church. I did the so-called Catechism classes there, but I would characterize those as almost agnostic. A corollary was my escape from any severe religious indoctrination in my childhood.

Andrew's son, my grandfather on my mother's side, the Reverend Frank Anderson D.D., made a career of that Baptist heritage. The Reverend Anderson and his wife Clara (Bergen), who was of Dutch descent, eventually lived in Des Moines, Iowa. They were warm and caring grandparents for me and my brother Dick. Their other daughter, my Aunt Leona, never had children. She often visited and became our dear Aunt. I had to rescue her in old age from a young swindler who robbed her of half of her life savings when she was living as a widow in Rosebud, Arkansas. That is another story, but I want to say that I needed help when I went down there to save her, and a couple from the small Baptist church there in Rosebud came forward to assist me. Together, we sent the swindler to prison, and managed Aunt Leona's final years until she died in a nursing home at age 90.

The Reverend Frank Anderson had originally married Clara's older sister Anne, but she died in childbirth. The baby boy died a month later. It must have been a sad time. However, his marriage to Clara a year later was a wonderful one as I remember them. She was always cheerful and bubbly. When she died in 1950, Reverend Frank more or less went to pieces.

The other quarter of my ancestry is Dutch. More correctly, those ancestors came from St. Jakob's Parish in Friesland. The summer after my second wife died in 1980, I took her son Patrick, 14, and my son Garth, 8, to Europe for two months, you might say to form a team of the three of us. One weekend we drove from our base in Bielefeld, Germany, over to St. Jakob's Parish and camped one night on Amelin Island there in the North Sea. It may be my imagination, but I saw many young men there with a strong resemblance to me: lean, blue-eyed, blond-haired, thin-faced guys. The name Bergen was changed from Kuiken by my great grandfather Arjen Kuiken when he emigrated from Friesland to America. When I travel in Europe and happen to be speaking French, invariably people ask me if I am Nierlandais (Dutch). I seem to look a bit more Dutch than Swedish, at least to Europeans.

On the other hand, my mother was a dark brunette, and my brother Dick is dark like her. Genes and ancestry are a funny business. There are rules of progression, but also outcomes that appear exceptional. I have always had the feeling that there is a lot of "latency" in our genes. And there is at times an underlying uncertainty about the identity of the true father.

* * *

In 1948, my parents announced that we were moving in June from Manchester to Boulder, Colorado. I was deeply upset. I loved Manchester and its river and my friends. Billy McCormick and I had built a kayak in his basement and I tried to talk him into running away with me downriver, much like Tom Sawyer and Huck Finn. But Billy had no reason to run. So at the end of my seventh grade, my brother Dick and I were in the backseat of a Buick headed west out of Manchester.

Fifty-five years later I went back to Manchester for the first time. The occasion was the high school Class of 1953 50th Reunion. Of course I also went to Boulder High School's Class of 1953 50th Reunion. But I always had warm feelings for Manchester and my childhood friends that I had to leave behind there. It was a good decision and I was warmly received, and saw John and Russell and Billy Post and Gordon Smith again (unfortunately, not Jackie). Jack Carr, brother of my classmate Alice Mae Carr – and carrying on the Carr family law firm – told me that he could still remember the Gustafson family "abandoning Manchester." When my family sold the Gambles store a few months before we left, the local newspaper ran a column, hoping such fine citizens would stay on in Manchester. Apparently, the whole town came out to wave goodbye to us, and sadly, as we drove away. Jack said I was known as one of the brightest kids in school – why were we leaving?

According to my brother, we left Manchester because my mother had an affair with Herb Schramm, and my mother and father hoped to rebuild their marriage in a new location. Herb had been my dad's assistant manager at the store before joining the Air Force and becoming a pilot. I suspect my brother was right, and I do remember that Herb, who was not married, did at times stay in our guest room when on leave. I really can't remember him after WW II ended in 1945. Was he fired? But perhaps the damage had been done.

We arrived in Boulder in 1948, and in short order my father started a successful insurance company. He also teamed with local banker Julius Kingdom and local lawyer Harl Douglas to buy the Mountain View Memorial Cemetery to save it from bankruptcy. Of course the marital objective of the move was futile. In December, 1950, my father divorced my mother and promptly ran off to Albuquerque, New Mexico with a new wife, Helen – a young woman closer to my age than his. He was 40, she about 22; my brother Dick was 18, and I was 15.

My father established a new insurance agency in Albuquerque, but in 1953 he moved to Seattle, where his older brother Gilbert lived. There, he started another hardware store called Luckies. Then he set up the new business curriculum at Seattle Community College. He would remain in Seattle until his death in 1988.

My brother Dick had graduated from Boulder High School in 1949 and enrolled as a freshman at the University of Colorado in Boulder. Thereupon he got into partying and flunked out. In the meantime, he had made pregnant the daughter of the Boulder High School principal, Owen Robinson. Lynn Robinson was a Boulder High senior at the time. Like his father, Dick had to get married. Shortly after, my father found him a job in a Gambles Hardware store in Alliance, Nebraska.

My mother rather quickly sold our Boulder house after the divorce, and she and I moved into a small trailer in the poor part of Boulder. The Gustafson family had completely disintegrated. My life's trajectory had bifurcated from that in a prosperous upper middle-class family to that in a single-parent family living in the Joratz trailer court, in what was the former red-light district of Boulder. I would learn the life of the poor and often carried a knife.

My father was furious that my mother has sold the house and moved us to a trailer. But what was she to do? The divorce had been bitter and he had given her the minimum. I never blamed her for the move to the trailer. I don't recall the change of status affecting my psyche negatively at all. I did, however, take "her side," and for many years thereafter my father and I were effectively estranged.

Later, I somewhat forgave him. He and Helen came to my wedding in Davenport, Iowa, about ten years later in 1961. Helen, while dancing with me, whispered that I should "bury the hatchet" with my dad. Shortly afterward they divorced, and he entered into a period of his life of heavy drinking. He would order cases of vodka to be delivered to his home. He showed up at my Ph.D. graduation in College Park, Maryland in 1965. I watched as he just took it as my mother bitterly attacked him at every opportunity. Between his third and fourth marriages (to the same woman, Betty), I invited him to visit us in Geneva in 1972, where I was spending a sabbatical year. One of his pieces of luggage had been lost in flight. We filled out the necessary forms and my father was asked by a Swiss official, "What should we look for inside?" My father gave a lame look and said there should be some shaving lotion on top. The bag was found a day later. It was one of those old doctor's satchels that opened on top. Under the shaving lotion were several cartons of cigarettes. Under those were several bottles of booze.

During his month in Europe, which he greatly enjoyed, he made contact with local Rotary Clubs, from which someone would take him to a liquor store. We put my newborn son in a *pouponnière* – a nice Swiss nursery to give new mothers a week or so rest from their child – and took my father on vacation to Ravenna on the beach in Italy. After about three days there, he and I were desperately searching for a liquor store. Wine and beer just would not do. I am sure that all that hard drinking and smoking shortened his life.

My brother Dick never forgave them. He once characterized our parents as "both spoiled rotten" – she, the pretty first child after her father's first wife and child had died in childbirth, and my dad the youngest son doted on by many big sisters. As I look back now, I agree with him. But my mother was very loving and in her final letters to us as I read them again, she cherished my brother Dick and me and our families. We were the most important thing in her world and I always felt that. She had to sacrifice a career as a concert pianist, and could not finish college; but he did not have to give up anything when they married.

Later on I was astonished when she went out to Seattle to help my father as he struggled with alcoholism between his second and third marriages. Of course he did not want her around. In personal matters, one must accept reality. However, her strict religious upbringing may have caused her to deny the reality of the divorce. Possibly, she also felt guilty.

My mother had always been a somewhat melancholy lady. She found a boyfriend, Jack Atkinson, a civil engineer who was working on the construction of the Boulder-Denver Turnpike. Jack was divorced with one child. He was a really nice guy and 50 years later one of my high school climbing buddies even remembered him and asked me what happened to him. Jack got me a summer job when I was 16 working on the Turnpike, even though I was two years underage. He and my mother were friends for many years and probably he loved her. But I think she always hoped to someday get my father back. My mother eventually went through a long depressed period and died relatively young at age 57.

I had to make a critical decision for her in 1958, just as I was near graduation from the University of Colorado. She was essentially in a deep nervous breakdown. My brother's family first tried to take care of her, but she ran away. Then good old Aunt Leona did her best for her sister, but to no avail. So my mother came back to Boulder where I got her to a number of doctors and psychiatrists. They all finally threw up their hands and told me she would have to be committed to the state mental institution in Denver. As she was decidedly suicidal and unstable, she was placed in the locked-down portion of the "mad-house." But I had refused to yet sign the papers to commit her. I will not forget the night I drove down there, under heavy pressure from the doctors to give consent. She and I talked by her bedside in the cell she'd been confined to. My mother begged me not to sign the papers. I did not sign. She got out and had ten more admittedly up and down years of life. Looking back, I am always proud of my decision that night.

* * *

Who really are we? I seem to be energetic and enterprising from my father's side. From my mother's side, I am idealistic and intellectual. Her father earned his Doctor of Divinity and was fluent in Hebrew and Latin. He was a crusader for women's rights and other causes. My Aunt Leona was Dean of Women at Drake University. I first, then my brother Dick, and later his son Phil, all have Ph.D.s in mathematics. Beyond the genetics, we are also a product of our environments. From Manchester, Iowa, I benefitted from a stable, simple, small-town childhood. Boulder later forced a number of character-building traits upon my personality.

Beyond our genetic makeup, which for simplicity I will take here as a causative factor in determining who we are, I will assert that who we are also depends on chance and even on choice. For example, suppose that the rich son of the local factory owner in Bystad, Sweden had not lusted after the innocent, delectable young maid? Or suppose my grandfather Gustafson had not developed a drinking problem when he retired back to Sweden? Or suppose the Reverend Frank Anderson's first wife had not died in childbirth? That's to say nothing of how my parents met to produce my brother Dick. And what were the events that ultimately caused each of them to be sent off to college at Carleton? What were their moods and needs that brought them to an act of love or lust one evening in Northfield, Minnesota?

So in June of 1948, by the choice of my parents I found myself in the back seat of a Buick headed west out of Manchester, Iowa, toward the new location of Boulder, Colorado, chosen largely by chance from among several other possibilities. My childhood in Manchester with all its myriad influences on who I am was being transformed to a boyhood in Boulder with all its very different influences.

Notes

Unlike the sad decline of many small Midwestern farm towns, Manchester survived and even grew from a population of 3,000 to 5,000 between 1950 and 2000. As county seat and site of the Delaware County Consolidated High School, Manchester certainly had some inherent advantages. My visit to the Class of 1953 50th reunion was my first return in the 55 years since we had left in 1948. To me, not much had really changed. I felt right at home. John and Russell and I marveled at our old haunts, which however did seem much smaller than in our memories. For example, the infamous Dry Run, a two-block system of storm sewer tunnels that we used to explore with great trepidation, looked really puny and harmless, and certainly not high enough to stand up in! I feel favored to have had such wonderful small-town values instilled in my childhood.

2. The Boy in Boulder

...and a lot of rock climbing...

One could barely make out the tiny figure moving slowly down the thin east face of the sandstone spire, which rose like a tooth out of the canyon south of Boulder. Very slowly, the climber descended in a straight line as deep dusk started to obscure all details. Then he seemed to disappear into a tiny depression on the right side of the face, about two-thirds of the way down. Silence reigned as darkness set in.

"Off rappel!" shattered the silence. A faint "okay" came from above. After a few moments, dim movements indicated the second climber was preparing his descent. "On rappel!" he called out, and he began to move even more slowly down the face. It was near darkness when he joined his climbing partner. Carefully they balanced on the edge of the face and looked down at their first ascent route on the north face, which they had put in and down which they must now descend.

The two climbers were Skip Greene and me. We were 16 years old. It was 1951 and we had headed south from Boulder to Eldorado Springs just to reconnoiter this spectacular pinnacle called The Matron that we had heard some CU rock climbers had their eye on. Skip had just bought his first car and we had driven it up what is now the South Mesa Trail toward Shadow Canyon from which The Matron thrusts up. We had a short climbing rope – my brand-new beautiful 300-foot rappel rope that I had obtained from Roy Holubar's little basement climbing supply store on Grandview Avenue in Boulder – and a few pitons and carabiners and very little sling rope. We circled the entire rock, negotiating the little cliffs and boulders around it, peering up at possible routes. When we came back to its north side, we saw what looked like the obvious route: A short steep layback, followed by some free climbing up to a cave under an overhang, and then up a crack to emerge onto the sloping east face.

Although it was already afternoon, we decided: why not try it? The route went, and quickly we were on the long thin east face. That was easy friction climbing although I can still feel its disconcerting tendency to push you over toward the drop-off down the massive south face. We were greatly disappointed to find a rappel wafer high up below the summit. We had thought we would have the first ascent of the whole rock. We needed that piton to get down.

"Pull the rope slowly. But when it comes loose up there, pull it fast!" Skip said. He was busy trying to pound a hole through a small lip of exfoliated rock on the very edge of the face. We had used up our last piton to rig the rappel from above. We had no hardware and no sling rope left. Our only way off the rock would be to thread our long thin rappel rope through that hole in the rock and hope that it would not break off.

"Right," I said, as I disengaged from my rappel and balanced carefully to start pulling one end of my 300-foot very thin quarter-inch nylon rappel rope from its anchor 150 feet above. That anchor had been our last piton, a little wafer, that I had been able to pound only a half inch into a tiny horizontal crack. Gently I pulled and even with its great elasticity, thankfully it started to come in as I let it accumulate at my feet. I could not see it above, but one knows instantly when the end goes through the little iron ring on the wafer and last 150 feet of rope goes into free fall. I did not want that portion of the rope to fall off over the south side of the face, where it would surely get hung up on some rock protuberances. Fortunately I was able to quickly bring the whole rope over to us as it fell from above down the face to my left even as I looked upward and could see nothing. "Got it!" I said. "How's that hole looking?" "I can hardly see what I'm doing!" said Skip. "But it's big enough for the rope to run through now." He kept pounding with his piton hammer, trying to dull the sharp sides of the jagged little hole about an inch or two from the little edge of the exfoliated rock.

"I don't like it." I said. "That's only a quarter-inch nylon rope. It can cut easily, and I doubt that we will be able to pull the rope down afterward." We discussed our options, and decided to use the sling rope off our piton hammers and also my rappel sling rope to rig a better anchor through the little hole in the rock. The rappel rope would then run through the smooth metal carabiner taken from my rappel sling. I would have to rappel the old-fashioned dulfersitz way, with the rope coming from behind between my legs then up across my chest and over the left shoulder to be held down behind with my right hand. For good measure, I grabbed my pocket knife to cut a piece off the bottom of my blue jeans and used that to pad the interface of our nylon sling anchor with the rock hole.

Skip went down first, putting as little weight as possible on the rope. Our first ascent route which we were now descending was steep but not vertical, so one could absorb some weight on tiny holds as one went down. "I'm down," he shouted from below. The anchor had held and as I had squinted at it, I saw no cutting of the ropes taking place. Very carefully, in total darkness, I eased myself over the edge onto the north face and went down. I still have the rope burn on my left shoulder from that rappel.

Our north face route is now the classic route up The Matron and was the second ascent of the rock.

From the moment that Buick from Iowa came over American Legion Hill on Arapahoe Avenue, where you first drop into Boulder Valley from the east, I have loved living in Boulder. Within a year, rock climbing would take over my life. By then I had fallen in with a group of young climbing friends. I really don't remember being much preoccupied by my family's disintegration during those first two years,

1948–1950, in Boulder. When one's life hangs in a delicate life and death balance somewhere on the rocks over Boulder, the mind is more naturally focused on footholds and handholds, how good the belay is, and what's the best next move. Mountain climbing, and especially rock climbing, has this Zen-like effect of cleaning and simplifying the mind.

My climbing began already at age 13 when I went over to my friend John Clark's house at 10th Street and College Avenue and his mother informed me that John and George Hall had decided to learn rock climbing and had gone up to the second Flatiron. I found them there and in our street shoes we climbed up that easy rock, trying to learn how to belay with the inch-diameter hemp rope George had just bought. Soon we discovered the Amphitheatre rocks in nearby Gregory Canyon, where we learned a lot of our rock climbing.

Eventually ten of us bonded together to spontaneously form what we called The Summit Club. The goal was to do some local rock climbing, and later when we obtained cars, to tackle some of Colorado's fourteeners (a fourteener is a mountain peak that exceeds 14,000 feet). In addition to John Clark, George Hall and me, there were the Vickery twins Jim and John, and their friends Cliff Chittim and Bill Fairchild. Corwin Simmons and Lynn Ridsdale joined us from University Hill Junior High. From Casey, the other junior high school in Boulder at that time, we added Skip Greene. Skip became my main climbing partner after 1951, when all of us went on to Boulder High School.

Most of the first ascents on the rocks around Boulder had already been taken by the older climbers at the University of Colorado, some of them ex-10th Mountain Division veterans, and other expert climbers such as Tom Hornbein who went on to the famed first traverse of Mt. Everest. We were not at all consumed with getting first ascents. But we did get some, just from our love of the climbing adventure.

There was The Matron North Face as recounted above. In addition, George Hall led the first ascent of the Amphitheatre West Pinnacle North Chimney, while we were still climbing in our basketball shoes. Jim, John, Cliff and I made the first ascent of the Schmoe's Nose high up on Green Mountain, by throwing a rope over a knob on its North Face and then climbing the rope hand over hand until you can stretch to your right to enter a short hand traverse. I was late to join this climb and when I got up there the others had just come off the rock. Undeterred, I threw my rope over the knob and climbed up it with no belay and was on top of the rock in a matter of minutes. I can still remember on my descent hanging off that hand traverse with my right hand, with no safety rope on me, and wondering if I could somehow get my left hand over to compress the two strands of the rope hanging off the knob. Then I could at least grab them and swing over to go hand over hand to the ground. Somehow I did it, but it was touch and go. John Clark and I climbed the south side of the Willy B. around 1950 after its first ascent by Hornbein, who put in a route up its east side in 1948. One spring afternoon in 1951 after school, the Vickery twins and I did the first ascent of the West Face of the First Flatiron, a crumbly rotten overhang. Basically Jim, who was as agile as a monkey, just pounded pitons in everywhere and John and I put our weights on the rope and tensioned Jim to the top. Then John and I hauled ourselves up, with Jim pulling from the top. We did not yet know the prusik technique (alternately sliding weight-bearing sling-loops up the rope), which would have been easier!

Two other climbing activities we initiated into Boulder, and which later spread into the worldwide climbing culture, were buildering and bouldering. Buildering means just what it implies: free climbing up the exterior of buildings. We introduced buildering first on Boulder High School and then extended it to climbing all that nice sandstone on the University of Colorado buildings. We put up first ascents on the Old Main and Macky Auditorium buildings. We started bouldering on a strange little rock just off the Mesa Trail, called Tomato Rock. Corwin Simmons deserves a lot of credit for then extending bouldering to lots of rocks on Flagstaff Mountain. One of those classic bouldering rocks is named for him. Among the more difficult buildering routes we put in was the Norlin Library south columns chimney. This is just the approximately 30 inch. wide space between those southern-most columns. Corwin and I made it to the top and down, of course without belay. When you finished wedging yourself up and down, your thighs were totally gone. If your leg strength would fail, you would crash down almost 40 feet between the columns to the entrance floor. I looked at it recently, and just shook my head.

Our innovations of buildering and bouldering in Boulder arguably presaged all the climbing gyms found worldwide today.

With the advent of our driver's licenses, we could venture farther afield. We became interested in Colorado's fourteeners and made winter or spring ascents of many of them over school vacations. The East Face of Long's Peak beckoned us and we climbed numerous routes there, including Stettner's Ledges and our 1953 first ascent of the Window South Corner at the edge of the Diamond. Because attempting the Diamond itself was forbidden by the Rocky Mountain National Park administration, in 1955 CU students Richard Becker and Jim Gorman and I put in the first ascent of the North Face of Mt. Meeker, which with Long's Peak completes the great cirque about Chasm Lake in Rocky Mountain National Park. CU student Bob Allen and I went over to Aspen in 1951 and made the first traverse of the Snowmass-Capitol Ridge, the last great unclimbed ridge between Colorado's fourteeners.

All of our climbing gear came either from the Army Surplus Store or from Roy Holubar's basement, from which he ran his mail-order climbing equipment business. Holubar's business later became a well known brick and mortar franchise. I claim to be Holubar's first employee (after his wife Alice, of course) as from 1949–1953 I would work down in his basement packaging things to be mailed, and get paid in pitons, ice axes, ropes, and the like.

Our parents exercised great fortitude in placing no restrictions on our climbing zest. Perhaps fate favored us by not killing or seriously injuring any of us. But we were very strong and agile on the rocks. Also we were fearless. And we had infinite endurance and energy.

A very serious situation which pushed us to our limits was our winter attempt on Mt. Lincoln during record setting low temperatures. The Vickery twins and Corwin Simmons and I drove to Alma and parked Jim's old Ford on the road south of Hoosier Pass and headed straight up Lincoln on snowshoes. We put in a camp at timberline as the temperature fell and fell. The Vickery twins were in one army surplus nylon tent, Corwin and I in the other. In our army surplus mummy bags we kept on all our clothes, even our parkas and gloves, with our boots inside with us. The cold became quite unbearable as the night progressed. We would beat on each other and call out to the other tent, "Jim, John, are you still there?" "Yes. Christ, it is so cold!" Dawn came and our little tents were shrouded in that frozen mist that occurs when a temperature goes below -40 degrees and the liquid phase has become impossible and there is only solid to gas transition. Our food was all frozen solid and our water canteens all cracked. The water had not leaked out, it was just frozen to the metal. We managed to get our boots on and stagger outside the tents. Corwin threw an orange at a rock and it shattered. Without a word we crammed our frozen tents as best we could into our packs and in what can only be described as a rout, we plunged downward through the deep snows to the car. Jim tried to heat the engine with his Primus stove but to no avail. A road grader came by and gave us a push to start the car by compression. Alma recorded a low temperature of -54 degrees. We would guess -60 degrees at our higher camp.

I probably saved Skip's life in his senior year at Boulder High. Skip had become enamored with a young summer botanist named Joyce, whom he had met at the research station. When she went back East, evidently she broke off the relationship. Skip decided to take his angst out on a solo attempt on the East Face in midwinter. Taciturn to the extreme, nonetheless he called me and told me his plan, and why. It would be, essentially, a do or die thing. However, if he was not back in two days, I was to come up to look for him. When he was not back in the afternoon of the second day, I became worried. I drove to the trailhead and post-holed through deep snow up to Jim's Grove at timberline and found his tent: empty. I went higher up to Chasm Lake and found nothing. I continued up a ways in bright sunshine but facing into a biting cold wind off the boulder-field at 12,000 feet. I decided to go no further, descended, returned to Boulder, and called Rocky Mountain Rescue. They found Skip at the little stone Agnes Vale refuge at 13,000 feet, freezing to death. He had climbed the East Face but had taken a fall and succumbed to exposure coming down the normal route and was not able to go any further. With hospitalization he was barely able to save his toes and fingers.

It is really quite extraordinary that ten teenagers self-organized into such a vital and healthy group and learned the deep comradeship and commitment necessary to survive in a life of extreme climbing. John Clark and I as high school students were asked to be instructors in the first Colorado Mountain Club Boulder climbing schools in the early 1950s. I was paid by the Boulder Chamber of Commerce to lead 600 (yes, 600) tourists up to the Arapahoe Glacier on their annual promote-Boulder glacier hike in 1952. Upon graduation from Boulder High School, our Summit Club just evaporated. Most of us went on to University, some did not. A few of us kept climbing but none with the same intensity. We celebrated that graduation in 1953 with a spontaneous decision to have a sleep-out on top of the third Flatiron. Eight of us spent the night up there snoozing precariously on little ledges near the top.

In those days, climbers like us were looked upon by our high school classmates as, well, pretty weird. Nonetheless, we integrated well into the usual school teen life of that era. Cliff Chittim was elected Head Boy of the Class of 1953. Jim Vickery was a star quarterback his sophomore year, before he broke his neck during a futile attempt during a game to win it single-handedly! He then became a cheerleader as had been his brother John. Corwin Simmons, despite his heavily muscular structure, was a sprinter on the track team, and I ran the quarter-mile. I also lettered in cross-country and was on and off the varsity basketball team all three years of high school, but I was just not aggressive enough to make first team. Also, I was elected Senior Class Vice President and given the 1953 Most Valuable Student Award, which is based upon a combination of grades and activities. But climbing was our main passion, and occupied our thoughts and free time.

Summers were spent working to make spending money. My first summer in Boulder I could do no more at age 13 than collect watercress from Boulder's irrigation ditches and grow strawberries in our back yard to sell to local grocery stores. The summer of 1949 found me rising at 3 a.m. and riding my bike down to the Watts-Hardy Dairy to be a loading assistant to the truck driver who made an early milk haul through Lyons and Estes Park, and even up to the Bear Lake Lodge in the middle of Rocky Mountain National Park. This job completely exhausted me and I have learned repeatedly throughout my life that I am neither an early morning person nor one who can go for extended periods without his sleep. In the summer of 1950, I worked turning lawn sprinklers on and off around Boulder. I also raked rocks at my father's Mountain View Memorial Cemetery.

My mother's boyfriend Jack Atkinson got me a job in the summer of 1951 in the preliminary Denver-Boulder Turnpike construction. Then I landed a job on the Trail Crew at Rocky Mountain National Park in the summer of 1952. In 1953, they made me a Foreman and gave me my own crew of eight. Of course I was a natural for these Trail Crew jobs and thought about becoming a Forest or Park Ranger. While in Estes Park we were called upon in several climbing accident rescues.

Where to attend college? Without even thinking about it I had finished third academically, behind two girls in the senior class of 200 students. I was top boy. I was encouraged to apply for the "full boat" Boettcher scholarship, but it indeed went to the top girl who was class valedictorian. Her name comes to me today, Doris Koerner, daughter of a CU professor. I was awarded a Regents Scholarship which would pay all my tuition and books and fees to any university in Colorado. Three of us were sent to be interviewed by the Harvard Club in Denver and scholarship offers were implied should we apply. None of us wanted to go to Harvard and be the smart but poor boys sent there to maintain the quality and reputation of the institution for the rich Eastern kids. I thought about the Colorado School of Mines in Golden, but decided I would be better off staying in Boulder where I knew the terrain.

Less than a week after our graduation in 1953, my mother bought me a second hand car, sold the trailer in which we had lived, and was on her way up to Buffalo, Wyoming, to buy a dude ranch and be close to my brother Dick and his young family. She had loaned him a substantial part of her divorce settlement so he could buy a small Gambles hardware store there. Both that investment and the dude ranch were not, in retrospect, wise financial decisions. But women, and especially mothers, often lead with their hearts rather than their heads. My brother, by working 14-hour days, made a break-even goal of it for a number of years before he returned to school to eventually become a mathematics professor like me. The dude ranch steadily lost money due to its short summer season, and after a few years my mother sold it.

Thus, June of 1953 saw me throwing away a lot of things so as to fit all my belongings into my car, and drive up to Estes Park for my Trail Crew job. Summer in Estes was fine. We did a lot of dancing at the Rock Inn, which was just five minutes from our National Park barracks west of Estes. It is rather amazing how youth has no worries. I had no idea where I would live in Boulder when I would start university in the fall. But three Korean War veterans on my trail crew invited me to live with them in a crummy basement apartment in Boulder as they too started as freshmen at CU.

* * *

Our house in Boulder 60 years ago was at 2432 Pennsylvania Avenue – at that time, a dirt road on the eastern edge of Boulder. Sixty years later I walk to work (from a university parking lot) in exactly the same steps as I used to take on my way to school at University Hill Junior High (now an elementary school). Our 1948 house did not last even 20 years and its space is now occupied by an Episcopal Church. Pennsylvania Avenue is now paved and called Colorado Avenue, and Boulder's growth has exploded eastward.

But the mountains are still there and always refresh me. In that walk 60 years ago, and today, I can see North Arapahoe Peak, named for the Indian tribe that historically wintered in the Boulder Valley. North Arapahoe Peak is a particularly beautiful mountain. Its configuration as seen from Boulder resembles that of Mt. Everest as seen from Rongbuk monastery in Tibet. Especially when it is cloaked in a new snow, I call it my mini-Everest. I have been up it a number of times.

I first climbed it with John Clark about 60 years ago. We chose to do so during an October storm which battered Boulder with 100 mph winds, even toppling a giant construction crane off the top of the power plant east of Boulder. Of course we did not anticipate that storm as we left Boulder. From the trailhead it's an easy hike to the top of South Arapahoe Peak, about 13,000 feet high. But then there is a short but significant ridge over to the North Peak which is also a bit higher. This ridge has steep gullies on its west side, and vertical drop-offs of about 1,000 feet to the Arapahoe glacier below on the east side. At some points you need to use your hands and it is definitely exposed.

We made it over to the North Peak before the storm fell upon us with both snow and an overpowering wind from the west. Going back along the ridge then became an effort at survival. I can still see both of us reduced to crawling, using our ice axes braced against little protuberances on the ridge to keep from getting blown off. The memory remains.

I last climbed North Arapahoe with my friend Kathy about 10 years ago. Kathy is an athlete although not a rock climber. The day was nice, and we moved steadily across the ridge. But at one very exposed point you had to climb a small vertical step. She looked between her legs at the glacier 1,000 feet below and started to freeze up. Standing with little purchase, she looked up at me and started trying with her one free hand to get her pack off to hand to me above. "No. Keep it on. Use both hands. Come on up." One should not play with balance at such points. She moved up. Only then did I give her a hand at the top.

I decide at this moment to take a break from this writing. I open the door to my balcony and step outside. My gaze shifts toward my left and the morning sun is striking The Matron which rises like a great fang where Bear Mountain descends toward Eldorado Canyon. My eyes slowly turn to the right and fix upon the Schmoe's Nose high up on Green Mountain. Looking northwest I see the summits of Long's Peak and Mt. Meeker rising above the foothills. I am home.

Notes

There has never been any previous written account of our unique Boulder High School climbing club (1951–1953) we called The Summit Club. Gerry Roach in his book *Transcendent Summits* (2004) claims to have originated the name The Summit Club when he was a Boulder High School student some years later. It could be coincidence but I have my doubts. Our names and climbs under that name were still in lore at Boulder High for several years. Ronnie Cox, Jonathan Hough, Dave Lewis and others a few years behind us certainly knew us and learned climbing from us under our banner of The Summit Club. The Colorado Mountain Club officially listed our Summit Club as supplying instructors for their Climbing School. Gerry arrived in Boulder from California in 1954 and started climbing in junior high school much as we had started.

At Roy Holubar's behest, I wrote a short account of our first-ascent traverse of the Snowmass-Capitol ridge, which appeared in *Trail and Timberline* No. 404, Colorado Mountain Club (1952), 119–121.

3. The Student in Poverty

...survival and success...

"Do you want to sleep in the bathroom, or in the coal bin?" Air Force veteran Tom Anderson asked Navy submarine veterans Doug McDonald and Ed Neil. Doug and Ed looked at each other and promptly replied in unison: "The coal bin, of course!" Tom turned to me, "Well, Karl, I guess we get the bathroom." Such was how the four of us began our student careers at the University of Colorado in the fall of 1953: three GI Bill Korean War veterans, and the youngster, me. The entry to the basement apartment was down three steps at the back of the house at 1012 15th Street, just a stone's throw from campus. A tiny kitchen with an old two-burner gas stove and no space for a table led to a larger room with four desks pressed against the walls, and a single small table in the center. At the far end a small door opened into a converted coal bin that had barely enough space for its two-tiered bunk beds against one wall. No windows. To the right was the other basement room, which held two small beds on an old concrete floor, along with a toilet behind a small partition and an open shower whose water just ran to the basement drain. The veterans had chosen this basement rat-hole for two reasons: It was very close to campus, and we could pay the rent.

In fact it worked well. True, I could see centipedes on the wall at night by my bed, and the kitchen was full of cockroaches. But Tom and I could just walk five minutes from campus and do homework in a single hour interval between classes. Doug and Ed would instead have coffee in the student union to ogle the coeds. I earned 17 hours of A's and B's and a 3.64 grade point average the first semester of Engineering School and knew I could make it. Tom, a Business major, did equally well. Doug and Ed did not do well.

My Summit Club fellow climber John Vickery, two years older than me, was in the ATO (Alpha Tau Omega) fraternity on campus. My brother Dick had pledged ATO. So I went through rush week and both fellow climber Cliff Chittim and I pledged ATO. I could hardly afford it but with no established family anywhere, I would use the ATO house and its attic as my base while at university. The ATO's were a good group of fellows, largely a combination of varsity athletes on the CU football and basketball teams, and some more wealthy out-of-state students who liked to drink and party. There were also the fellow climbers Bill Pugh and Bill

Bueler. I would share a room one year with Pugh at the ATO house. With Bueler I would share 50 years of mountaineering adventures.

The first such adventure came when Bill invited me to join him and his friend Harvey Carter from Colorado Springs on a trip during Christmas vacation in 1953, to try to climb Mexico's high volcanoes. Harvey was an ex-mountain trooper and went on to become a rock climbing legend, pounding pitons up obscure first ascents all over the state of Colorado. We had no car so we took Greyhound buses down to San Antonio, Texas, sleeping in our sleeping bags on the floor in the aisle. Harvey had a friend in San Antonio who had an old station wagon to drive down to Mexico City. He would not climb, just wanted to see the place. We had numerous mechanical problems with his vehicle but somehow, driving night and day, and pounding on the carburetor from time to time, we got it to Mexico City where we had to wait several days, camping in someone's front yard, for some blond German auto mechanics to get it working again. This was my first experience outside of the English language culture of the United States and Canada, and my one year of high school Spanish put me in the position of translator. Finally we drove to the 13,000-foot saddle between the volcanoes: Ixtaccihuatl (17,343′) and Popocatepetl (17,887′), and after an evening cooking and snoozing in the large open refuge there, we climbed Popo the next day. Harvey had a terrible time with the altitude but willed himself to the top, retching several times along the way.

Then we drove over to climb Pico di Orizaba, the highest of the volcanoes at 18,700 feet. In those days there was no climbers' hotel, so we rented mules to take us up to about 15,000 feet, where we spent the night out in the open in our sleeping bags. It was cold. But I still remember the glorious sight at dawn of the sea of clouds below us. Again, Harvey retched his way to the top. Bill and I, on the other hand, reveled in the experience of a foreign culture.

* * *

Bill Bueler became one of my most trusted friends, and I his. Upon his graduation from CU he did three years service in counterintelligence for the U.S. Army, where he began to learn Chinese. Thereafter he worked for the CIA in Asia for seven years, during which he perfected his Mandarin Chinese language skills. His main task was the debriefing of Chinese spies as they came out of the mainland. They wanted to talk to "a real American" – that is, not a Chinese-American from California. Bill was about 6′ 4″ and moreover possessed a natural mild, non-threatening personality. Because of his mountaineering skills, Bill also played a key role in bringing the Tibetan Khampa guerillas from Tibet to Camp Hale in the high Colorado mountains, where they were trained in paramilitary skills and then sent back, most likely to die, to Tibet to harass the Chinese occupation there. Because he disagreed with the United States policy toward China and Tibet, he eventually resigned from the CIA.

Coming back to Boulder in the 1970s to possibly pursue a Ph.D. in Chinese Studies, Bill accompanied me as I completed the ascents of all Colorado's 14,000-foot-plus mountains. His knowledge of China and his Chinese language skills were superior to those of faculty here, so Bill did not fit in. Washing dishes and caring for his two small children, Bill adopted a secondary role as his wife Lois finished her Ph.D. in Comparative Literature. They moved to California where she became a professor at Chico State, and Bill became an Instructor of Chinese at the Defense Language Institute in Monterey. In 1994 Norm Nesbit and I would join Bill for a fascinating trip through China into Tibet.

In June 1989, Bill and Lois happened to be in Tianamen Square on the very day of the infamous incident. Most foreigners immediately fled China, but Bill and Lois, with the help of the Chinese Mountaineering Association, were able to get on a train and travel west across China to Urumchi to complete their trip as originally scheduled. He gave Norm and me a copy of an incredibly interesting written account of all his conversations with the Chinese people on that long train ride out of Beijing, an account he said he had to write "for the agency."

All told, as China opened up, Bill made thirteen trips to the remote regions of China and Tibet. Was Bill still a deep-cover agent for the C.I.A.? We can only surmise. Certainly Bill's demeanor, which was like that of a harmless college professor, was quite disarming. But my colleagues at Berkeley tell me that the DLI in Monterey was a vital reporting link through which its language trainees came and went on their intelligence missions.

In a short private autobiography he wrote for his family shortly before his death from a brain tumor in 2004, Bill stated that Norm and I were his two best friends. He wrote that he "could only imagine" my rock climbing skills. That was an exaggeration because Bill never did any serious rock climbing. Instead, his dream was always to summit mountains. Bill was his happiest as he munched on a lunch on top of a Colorado 14,000er. I enjoyed many such lunches with this earnest, intelligent, humanitarian dreamer.

* * *

In the spring semester of 1954, my three roommates and I continued to scrounge to get by in Boulder. We would all go down to the Elks Lodge one night a week and serve as kitchen help and waiters. The real payoff, beyond the money, was the huge pots of leftovers the Elks would let us take home. I can still taste the fabulous Swiss steak we would eat for days. I was beginning to learn how much it cost just to live; in particular, just to eat. A solution was to be a fraternity and sorority hasher, i.e., kitchen help or dining room waiter, throughout my college career. In so doing, one accessed 20 meals a week, with the time lost from studies largely clustered around meal-times. I was head hasher at the ATO house for several semesters, gaining both room and board.

"Let's make some big money up in Alaska!" Denny said. Denny Marriott was an ATO a year older than me and would eventually join the Marines. So when the spring semester ended, he and I earned free passage to Seattle by driving so called "drive-away" used cars west from Denver for a used car dealer. Our plan was to free-load for a couple of days off my father and his wife Helen and figure out some way to get up to Alaska. I think we left Boulder with about $25 each. But Denny infuriated my father the first night at dinner. I have forgotten the topic but Denny was losing an argument with my father, who after all, in college had been a champion debater.

Denny: "Well, Mr. Gustafson, the reason you can't see my point is because 'old dogs' can't learn new tricks'."

Dad: "Look, young man, you just better watch it, you are here enjoying our hospitality, our food ···"

Denny: "I think there should be an adjustment for age on I.Q. tests, then you would know you are slipping, Mr. Gustafson."

Dad: "You two are out of here in the morning!"

The next morning in a drizzle Denny and I shouldered our duffle bags and hitchhiked north to the Canadian border. The Canadian official looked into the car and asked the driver, "How far down the road did you pick these guys up?" We got out and our duffle bags were searched. The official pulled out Denny's 45 caliber pistol and said, "Don't you know you can't bring that across the border?" Denny and I hitched our way back to Bellingham and slept beside the road somewhere. We bought a breakfast and hatched a new plan. First Denny mailed his weapon back to his Aunt. Then we consulted boat schedules leaving Vancouver toward Alaska. We fixed our story between us and bought bus tickets to Vancouver. At the border we were taken off the bus with all others and separated. Our story held and we were back on the bus and in Canada. Short of Vancouver we got off the bus and hitched or walked over to the highway heading north up the Frasier River. No one picked us up and we slept in the rain in the borrow pit by the road. The next morning we shaved and looked better, and got our first ride north. That night we were abandoned in a wild region but the rain had stopped and I remember our walking for miles through the night with our duffle bags alternated from shoulder to shoulder until we got a ride in the morning. We continued to get rides north and ate little. I think it was near Prince George that some grandfather with his teenage granddaughter picked us up. Denny and I just looked at each other and marveled at the old fellow's trust. But we always kept ourselves clean-shaven wearing reasonably clean shirts and blue jeans. In Dawson Creek we stole soup and cans of food from a grocery store and ate them cold. Finally a United States Air Force Lieutenant driving alone to Fairbanks picked us up and said we could sleep on the floor of his motel rooms if we would help him with the driving. In those days, the Alcan Highway was largely unimproved gravel and driving was tiring. Denny and I would buy a large pancake breakfast that would have to suffice for the whole days' nourishment. Finally we arrived in Fairbanks, starving and dead broke. We took the first job we could find.

That job was at the large open-pit gold mining operation called Fairbanks Explorations, southwest of Fairbanks near Nenana. The very high post-World War II salaries that had lured us to Alaska had largely vanished. But we needed food. The FE gold mining operation provided a dormitory to sleep in, three superb worker's meals a day, a seven-day work week, and even two shifts a day if you wanted them. We wanted them. We made a lot of money over the summer, and spent little, going into Fairbanks only once a month.

I really liked it there. Usually I was alone at a conveyor belt hopper station, through which passed huge volumes of soil, as about one hundred feet of the earth's surface was unfrozen and removed by a huge dragline to get at the layer of gold beneath. The gold would then be dredged out in an artificial lake constructed there.

Sometimes Denny or I would ride the belt to chat at each other's station when the earth was flowing smoothly through our hoppers. When the earth was too wet, it was a frantic nightmare as the hopper wanted to clog. I would shovel and push desperately and exhaust myself during the entire eight-hour shift. When I could not keep up and the hopper started spewing out earth over all its sides, there was a red button to turn off the whole conveyor system.

We started smoking to keep the mosquitoes at bay. Then the black flies came. There are a lot of bugs in Alaska. My arms are still scarred as testament to some monster mosquito bites.

As fall approached, I did not want to leave the two little foxes who would come out to eat with me when I worked swing shift (3 to 11 p.m.), my favorite shift. And I needed as much money as possible. The six of us college students among the 1,000 Eskimos, Indians, Finns, and Swedes who worked there, delayed our return to the lower 48 as long as possible. Finally we piled into one car and headed south. Near Whitehorse at night the driver fell asleep and I remember a rude awakening in the back seat as the car plowed into the ditch. Pulled out by mules and then towed to Whitehorse cost us two days waiting for repairs. From Montana, Denny and I hitched down to Boulder. We were two weeks late for the semester. Nonetheless I earned 14 hours of A's and B's.

That's how my five years of college went. I struggled to get enough food to eat during the semesters, and spent summers trying to make money. One summer Dick Bird and I flew up to Juneau, Alaska, looking for work and took the only job we could get, working on a survey crew deep in the swamps east of Fairbanks. The mosquitoes were so bad you ate your lunch with a mosquito net around your head. You were in water, sometimes up to your neck, all day. One evening I returned to the little boat in which the crew went up the river and jumped from the river bank down to it. At the other end of the boat was a large black bear helping himself to our provisions. The story is that the boat scarcely sent out a ripple as the bear and I simultaneously jumped back up to the bank above the boat.

Another summer I worked for Shell Oil on two oil prospecting boats that would head out of Galveston, Texas, to carry out seismic surveys of the layers below the Gulf of Mexico. One boat trailed a long cable of recorders while the other boat would come to the center of that cable, and we would throw out a dynamite stick to create the sound wave that would then reflect the layers back up to the recorders.

One summer I found work in Boulder so that I could catch up on my course credits by going to summer school, for I needed 180 credit hours to obtain the combined Engineering Physics – Business Finance B.S. degrees.

Recently I obtained a copy of my CU transcript, my own having been lost long ago. In spite of all my outside employment, I finished those five years with a 3.084 grade point average. In those days that was rather high. I ranked 91/335 and in the first quartile in Engineering. In Business School I was 37/234. In Applied Mathematics, I was 1/5. I was inducted into the two Engineering Scholastic Honaries Tau Beta Pi and Sigma Tau. I was elected President of the local student chapter of the Physics Honorary Sigma Pi Sigma.

It was while I was president of the physics honorary that the world changed forever. On October 4, 1957, the Soviet Union took humanity into the Space Age with their launch of Sputnik, the first artificial satellite to orbit the Earth. I remember specifically the night I first saw it. As the president of the physics honorary, I had to do all the work and that was not much, the main duty in the fall semester being to organize the banquet. This would be held at what I recall was then Blanchard's Lodge (now the Red Lion Inn) in Boulder Canyon, a few miles west of Boulder. For dinner speaker I had sought the famous physics professor George Gamow, but I could never find him. So somewhat in desperation, I asked Professor Carl McGuire, who was teaching the International Trade course I was taking, if he would come tell us about his time at the American University in Lebanon. McGuire was a noted authority and in the Economics Department and of genial disposition. The latter attribute was fortunate because just as McGuire stood up to start his after-dinner lecture, Albert Bartlett, the Sigma Pi Sigma physics faculty adviser, looked at his watch and said, "It should be going over now!" We all ran outside and waited to see Sputnik cross the heavens, tracing its dramatic arc across the night sky. McGuire came out too and enjoyed the event. I see from my transcript that I still got a B in his course, certainly all I deserved!

My enrollment into Engineering School at CU in 1953 effectively ended my days as an expert rock climber. Between the workaholic nature of the engineering students and my need to work for meals and also at any other small jobs, I had no time at all for climbing. In particular, I missed out on the opening of the now famous Eldorado Canyon rock climbing mecca. I was invited to join the team of Dick Bird, Cary Houston, Dale Johnson, Chuck Murley, and Dallas Jackson, all of whom I knew, on their pioneering 1956 Redgarden route, which required several Saturdays of climbing.

In the winter of 1955–1956 I destroyed my left knee skiing at Arapahoe Basin. A Boulder doctor who volunteered his time to moneyless CU students, Dr. Maxwell, found some remote source of funds to operate and repair my ligaments. Exactly 20 years later I further destroyed that knee playing pick-up basketball at CU with graduate students. That time I had insurance and Dr. Maxwell was again the surgeon, this time removing the meniscus. I woke up in the recovery room and felt virtually no pain at all. He stopped by my hospital room that evening, and we concluded that he had cut through into the knee in such an automatic way exactly as in the previous operation, so no new nerves had been touched. He then offered to immediately discharge me from the hospital. But I had five children at home in my new combined family with Rose and I knew I would be prone to re-injuring the knee. So I demurred. With great sense of humor, Dr. Maxwell asked me if I was more afraid of re-injury or facing the care of five kids; then he asked me how many nights I wanted at the hospital. "How about two, Dr. Maxwell?" "Done, Karl."

As a result of my 1956 first knee operation, I decided I could no longer count on it to be 100 percent, and I vowed thereupon to never take on any difficult rock-climbing leads again.

Nor did I have the money or time for any girlfriends in my undergraduate years of 1953–1958 at CU. If some sorority girls would buy the pitcher of beer, we would go dancing on Friday evenings at Tulagi's or Timber Tavern. A lovely girl named Muffy French liked me but I don't know what happened there. Many of my fraternity brothers spent a lot of money taking girls down to Denver, but I never did – and to this day, I don't! I have always regarded Denver as a long trip, and too expensive! Later in my life, one of my lady friends who moved from Boulder to Denver for work somewhat frustratedly told me, "Karl, the only time you come to Denver is on your way to the airport to go to Brussels!" She was right.

As the Cold War between the Soviet Union and the United States intensified in the 1950s, the United States belatedly started giving more priority to its higher education system. I benefited from that trend when the College of Engineering instituted its Engineering Problems Instructors. Twelve undergraduates were selected to teach the one credit hour course in slide rule and mechanical calculator required of all Engineering first year students. I was selected from Engineering Physics, others came from the Electrical, Chemical, Mechanical, Aeronautical, and Civil Engineering majors. We were all put in a room in the basement of the Ketchum Engineering building. It was a great experience. Dave Clare went on to a vice presidency in Exxon Oil. Ray Essert went to IBM. I suppose that was my first experience of being selected into an elite. Certainly my three semesters of undergraduate EPI teaching in 1957–1958 were the catalyst to my entering a lifelong academic career. And I could use the money.

The Department of Applied Mathematics had obtained permission to grant a Bachelor of Science degree for the first time in June, 1958. I had taken so many applied mathematics courses that it was quite easy to convert my official primary major from Engineering Physics to Applied Mathematics. Five of us were so recruited from the engineering disciplines and constituted the first B.S. recipients in Applied Mathematics from the University of Colorado. I was offered a full-time Instructorship in the department and I took it. I bought a used Chevy convertible and felt secure financially for the first time in many years.

In that era, many engineering graduates were encouraged to go on to law school and thereafter into the lucrative profession of patent law. I moved into a house at 969 12th Street, to share one of the two bedrooms with another instructor of Applied Mathematics, Fred Rumford. Fred wanted to go to law school, so we both enrolled part-time for the coming academic year 1958–1959. I was not really motivated to become a patent lawyer, nor was Fred – but we thought law might be interesting. The Dean at the CU Law School had encouraged me when I had discussed it with him, and as I recall, I scored the highest of all applicants from Boulder on that year's LSAT. We both found the fundamental first-year course on Contracts very fascinating as taught by the legendary Professor Stork, who would enter the classroom, sit down, put a heating pad over his legs, and start rapid-fire questioning about the cases that had been assigned for the day. The game was: you better be ready, if it was your day to be called upon! You only had one or two chances per semester. When your turn came, you would be engaged in the Socratic method with Professor Stork for up to 15 minutes.

Fred and I were, however, definitely an oddity among the more traditional law students. Almost all of them were full-time students, putting in 12-hour study days in the law library. To keep short here the law school story: Fred and I survived. Both semesters, we crammed hard together for the final exams. I see from my transcript that I got B's both semesters in Contracts, and a C in Property, which I found incredibly boring. As I recall Fred got all C's. We decided we would rather be mathematicians than lawyers and enrolled as graduate students in the (Pure) Mathematics Department for the coming year 1959–1960.

Fall 1959 thus found us both in the course in Real Analysis, taught by the renowned Professor Wolfgang Thron. The stories about Thron giving low course grades were legendary. Here is one. His Ph.D. student Arne Magnus was teaching a seminar joint with Thron. At the end of the semester, Magnus found that Thron had given him a C for the course. A mild-mannered man, nonetheless Magnus went to see Thron about it.

"How can you give me a C? I gave almost half of the lectures!"

"Vell, I vould only grade myself B in the course, and I know more about the subject than you do!" Thron replied.

This was my very first pure mathematics course. We first went through the classic Grundlagen der Analysis by Landau, a grueling text with few words in which one derives the properties of the real numbers from just a few axioms and arguments with inequalities. Then we used a more reasonable book on the more advanced aspects of calculus, by Rudin. I received a B, Fred a C as I recall. Fred had come out from Kansas to learn to ski; I had some money now and also wanted to do some skiing. So we decided to do more skiing the spring term and we did not enroll for the second semester of Thron's course. Between semesters I drove over to Aspen and indeed did a lot of wonderful skiing with my climbing friends Dick Bird and Harvey Carter who were on the Ski Patrol there. Aspen was fun in those days and one could sleep on the floor at a friend's place, have a beer at the Hotel Jerome, and hear Glenn Yarbrough sing at the Limelite, all on a small budget.

But as the spring semester's teaching started, the secretary in the Applied Mathematics Department called me and said Professor Thron wanted me to come over to his office in Hellems where the (pure) mathematics department was housed. I dutifully went over.

"Vhy aren't you taking the second semester?"

"Well, I only got a B in your first semester."

"Vell, that vas the top grade in the course. You have more talent than most of the full professors in that applied department. Vhy don't you sign up for second semester?"

I did. I received a grade of B again. I was on my way to a Ph.D. in Mathematics. But my life would first be dramatically interrupted for a few years.

Notes

Curious as to whether the infamous 1012 15th Street basement apartment might have changed, or was even still there, I walked over to peek, now 56 years later. No change at all! The little house is still partitioned into four small living quarters, the best one upstairs, ours below, and two other tiny cubby-hole doors in the back next to our entrance. I did not inquire within.

4. Computers and Espionage

...and the world's first spy satellite...

It was 1959 and the Cold War was escalating steadily, moving from a state of palpable sustained tension toward the overt threat to global peace to be posed by the 1962 Cuban Missile Crisis – the closest the world has ever come to nuclear war. Quite by chance, I found myself thrust into this vortex, involved in top-level espionage work. I would soon write the software for the world's first spy satellite.

It was a summer romance, in fact, that that led me unwittingly to this particular role in history. In 1958 I had fallen for a stunning young woman from the Washington, D.C., area, who had come out to Boulder for summer school. So while the world was consumed by the escalating political and ideological tensions, nuclear arms competition, and Space Race, I was increasingly consumed by thoughts of Phyllis. We were still infatuated with each other by 1959.

So in late spring of 1959 I decided to look for a summer job in Washington. From a list of government agencies offering scientific summer internships, I selected one at random and made my first telephone call, to a place called The Naval Research Laboratory (NRL). A rather bureaucratic-sounding voice answered. He identified himself as Bruce Wald. "Do you know any computing?" "Yes," I answered. I had taken one course out of a book and had never touched a computer. Be assured, there were very few computers around in 1959. "What were your college majors?" "Engineering, physics, and mathematics," I replied. Pause, "ahem," another pause, a throat clearing, then, "Okay, well...we would love to have you come to work with us here. You have the job. Get here as fast as you can."

The day after teaching ended I drove straight through to New York City. I had bought a better (used) car, a Chevy convertible, and a girl from the Chi Omega Sorority going home to Scarsdale, New York, paid for the gas. After a quick night's sleep on a sofa at her parent's house, I drove down to Washington. My girlfriend Phyllis lived in Chevy Chase at the north end of the District of Columbia diamond, NRL being at the extreme southern tip of the D.C. diamond. Probably I stopped to see Phyllis but I am sure I also went promptly down to NRL that first day in Washington. I do know that Bruce and I went to work on his LGP-30 computer the very first day I was there. Immediately the papers for a Top Secret clearance were initiated for me. Things were moving fast. I was introduced to key personnel under whose supervision I would be working. Two of them were Bob Misner, who seemed to have endless larger than refrigerator-sized magnetic tape recorders standing in rows throughout the building. Another was Mack Sheets, who seemed to be an expert on something called the Wullenweber, some kind of German device for determining the angular bearing of an incoming electronic signal. I was in the Intercept and Data Handling section of the Countermeasures Branch of the Radio Division at NRL. The branch head was Howard Lorenzen, one of the world's experts and a veteran in the game of electronic warfare. Lorenzen had contacts in and was an integral member of the highest Intelligence circles within the United States government.

Although it would take 18 months for my security clearance (above Top Secret) to be completed formally, no one wasted any time bringing me up to speed on the primary task: locating Russian submarines from their electronic transmissions. The method is called direction finding. From listening stations arrayed around an ocean, you triangulate the submarine's position. Even a one second coded burst to Moscow can reveal a submarine's position. Bruce's task, and now mine, was to get rid of all those full-wall ocean maps with British girls standing on ladders drawing bearing lines across them (that you see in the old World War II movies), and put the whole calculation into our little LPG-30 computer. If we could succeed, it was clear that the Navy would give us an essentially unlimited budget to issue contracts to industry to develop larger scale systems for general ocean surveillance in both the Atlantic and Pacific. During the summer, I soon surmised from the locations of some listening stations surrounding the Soviet Union that our computerized system would also be used for surveillance over land areas as well. Not only the Soviet Union but also Communist China would be under surveillance. I learned later that these high frequency direction finding (HFDF is the acronym) systems had been used in 1957 to track the Soviet Sputnik from its 20 MHz signal and thereby determine its orbit.

Direction finding is a really fascinating weapon within the panoply of electronic warfare. Unlike radar, you send out nothing. You just listen. No one but you needs to know your location. In that era and even today it is not widely known to the public, and its specifics are closely guarded secrets. I had by romantic accident been brought into some of the world's most secretive intelligence projects.

NRL had rented an apartment for me close to the laboratory. Bruce and I were given unlimited overtime and often worked late into the evenings on our little computer. Other evenings I would commute up Rock Creek Parkway to see Phyllis, who lived near Chevy Chase Circle. Sometimes the maid at her parents' house would make me breakfast and criticize me for not having an apartment nearby. Sometimes Phyllis came down to my apartment in Anacostia, but not often. Increasingly, I found myself unable to fit into the wealthy Catholic culture of her family. On the other hand, I could not wait to get back to work on the fascinating tasks the Naval Intelligence community had laid in my lap. Moreover I knew by then that NSA was the real client for much of the information in Bob Misner's large banks of magnetic recorders which were being fed daily from intercept networks.

The lay public may not fully appreciate the value of knowing the total electronic signature of your adversary's military installations and devices. We humans think of espionage information as what we can see with our eyes or hear in our ears. But the electronic spectrum is actually more valuable if you wish to know your adversary's capabilities and how to knock them out. Compare it to boxing in a ring. It is necessary to see and even hear your adversary. But it is of higher value to determine his strengths and weaknesses, where and when to be able to strike, or counterpunch.

Essentially, nobody knew computing in 1959. Of course there was Los Alamos, and a few academic centers that were forming computer staff. There were no Departments of Computer Science at universities. Certainly none of the 50 or so radar engineers in the Countermeasures Branch knew any computing at all. Mature electrical engineers like Bob Misner and Mack Sheets who were the immediate superiors to Bruce and me really had no choice but to always try to meet our demands. We would take two hour pizza lunches but then work almost all night. I was pretty harmless as a young protégé but Bruce enjoyed pushing the limits and taunting his superiors with our work-time wishes.

Of course I had no idea to whom I was talking when Bruce Wald hired me sight unseen. When I arrived in Washington, D.C., for the first time at NRL's heavy-security front gate, they parked my car and I waited to meet Bruce. Pretty soon he arrived. I won't say he waddled up, but he was pudgy and walked with a stride of toes-pointed-out. It was my first encounter with a young genius. My experience at Colorado had not had any talk of genius among us. We were all just bright hard-working engineering students.

Bruce had graduated from Bowdoin College in Maine at age 18 and came directly to NRL. I don't know where Bruce learned computing. Maybe he just learned it himself. I remember one day later on Bruce had loosely wired some transistors together and was applying various voltages to his little makeshift circuit. He was just holding them in his hand. Transistors were just coming on-line.

I suppose, in a way, that Bruce was as curious about me as I was about him. Later on he would refer to me as "that extra-smart ski-bum from Colorado." Certainly I did not fit the Eastern United States image of genius, geek, nerd, or intellectual. Bruce did. But we got along very well, although we never became really close friends. We did have hiking as a common denominator. Bruce and I and my girlfriend Phyllis took a long weekend to drive to the White Mountains in New Hampshire and did a two night hut trip over Mount Washington and Mount Madison. Two years later in 1961, when I had a month between my Basic Training at Fort Knox, Kentucky, and my marriage to Becky Emeis in Davenport, Iowa, in June, Bruce and his wife Betz put me up at their house some miles south of NRL on the peninsula. That was indeed an experience. I knew that Bruce just let trash accumulate, indeed, pile up in his car. He would finally take it to the gas station and pay an attendant to just haul and vacuum it all away. Their house was in about the same condition. Eating meals there was trying. We ate at the kitchen table across from a cage they had installed in the kitchen with a couple of monkeys in it.

One would just throw scraps from the table over to the cage. Sometimes the monkeys threw them back.

Like many brilliant scientists, Bruce rebelled against seemingly senseless rules. But our supervisors liked it when we volunteered to work on the computer all night. Go ahead and fill out your own time cards, they said. I see now how they needed us desperately. I see now how the nation needed us. I could have had no better teacher of pioneer computing than Bruce Wald. With our countervailing personalities, we played off each other, challenged each other. By the end of the summer, we were computing experts.

I shall always be grateful that I learned all of the basics of computing on such a primitive (by today's standards) computer. One started the LGP-30 with a small punched paper five-hole tape boot-strap program, which would then pull in the longer tape containing the machine language instructions we had coded. The arithmetic was in base 16 (hexadecimal). I could "one-op" the program, looking for bugs (coding errors), as all bits being processed could be seen sequentially on small registers at the top front of this desk-sized computer. I grew to love this little machine. I knew where every bit came from, and where it would go. Apparently there is a computer museum near Stuttgart, Germany which maintains a working LGP-30.

Fascinating as this pioneer computing and intelligence work was, with the cooling of my romance, and the almost intolerable Washington summer humidity, I started to dream about getting back to God's country – Colorado. I wanted to be in my comfort zone of Boulder and start graduate school toward a Ph.D. in mathematics. Looking back, I'm somewhat amazed at how I decided to just dump that work so vital to the United States intelligence community and head back home. The Navy offered me strong incentives to stay on. I could go to graduate school, anticipate a rapid rise in the GS salary system, the like. But the pull of my mountain kingdom was stronger; and the end of summer 1959 found me happily driving back to Boulder.

Not relenting, the Navy contacted the head of the Applied Mathematics Department, Charles Hutchinson. He called me into his office and said the U.S. Government wanted him to create a secure safe room and security safes for me to continue the military work. They would even pay for security guards. Hutch (as he was called) was a craggy old fundamentalist and he was simply incredulous. I still remember him looking at me with a mixture of bafflement and not a little resentment. How could this young pup of an Instructor have become so important? The University of Colorado at that time had no classified research capabilities. Of course I could not tell Hutch the specific intelligence task to be undertaken. The department had just acquired a small computer, a Bendix G-15, which was comparable to the LGP-30. But a moment's thought told me that I did not want to take on the onerous and time consuming machine-coding all over again on a different machine. And I wanted to start taking pure mathematics courses. Quickly putting it all together, I took a deep breath and told a relieved Professor Hutchinson, "Tell them we just cannot do it."

As recounted in the previous chapter, the 1959–1960 academic year found me successfully teaching applied mathematics, learning pure mathematics, and starting to enjoy life as a ski-bum. Then out of the blue in the spring of 1960, I received notice from the Selective Service System that I was drafted. I was told to report for a physical exam and expect to serve four years in the U.S. Army. I could not believe it. Other university instructors my age were not being drafted. I asked Hutch if he would sign papers categorizing me as a critical skill, exempt from the draft due to national teaching needs. He demurred. I protested. He suggested that if I didn't like his decision, why didn't I try my colleagues at Naval Research? I consulted a Mr. Rhodes, the veteran's affairs advisor on campus. He agreed with Hutch: if Hutch would not sign, the only way out was if the Navy would put through the critical skilled enlistment papers. I would have to go back to Washington to work for them for four years, do a three-month Army basic training, and be in the Army Reserve for eight years. The likelihood of being called up for active duty in case of war still existed, but the probability of that was reduced by the fact that I would not have the usual additional three-months training for a specialty designation in the Army. I wouldn't be assigned to any specific Active Reserve Unit, and instead would be placed in the Inactive Army Reserves.

After a few days' thought, it seemed pretty clear. Graduate School and skiing would now have to wait. And I knew the intelligence and computer work would be fantastic. Another positive factor was that if I got my Ph.D. back East, e.g., at the University of Maryland, I might be more able to come back to Colorado permanently as professor. It is generally considered a sin for a University to hire its own Ph.D.'s, although they do it sometimes.

Thus I called my friends in the Countermeasures Branch at NRL. They were delighted and immediately started the paperwork. I was told to report immediately upon the end of the teaching semester. One has to wonder if any backroom dealing had taken place. One can never really know in the spook business.

My future wife Becky went back to her parents' home in Iowa, to work and wait until I completed the Army Basic Training, after which we would marry. I went back to Washington and even deeper into the electronic espionage game. Bruce and I successfully demonstrated our prototype codes on the LGP-30 for the submarine position fixing tasks. We were elevated to scientific officers to supervise multi-million dollar contracts to industry to build large state-of-the-art computers to be put into the military operations centers. I was placed in charge of all mathematics and software development. An electrical engineer, John Ihnat, was made responsible for all hardware aspects. John remained on this and related projects for a lifetime career at NRL. Bruce later rose to an Associate Research Director at the Naval Research Laboratory.

I have detailed elsewhere (*see* Notes) some of the mathematical tasks and scientific challenges facing me as I found myself at the forefront of this wave of computer innovation during the following four years. Bruce, John and I chose Burroughs Corporation over IBM for the contract to build the new computers. Working closely with Burroughs, we participated in the rapid hardware progression from magnetic drum to core to solid-state memories. We favored and chose Algol over Fortran for the software language and compiler. Our resulting military computer was called the D-825, and some of our ideas later appeared in the Burroughs

B-5000 and Illiac machines. We created the computing concepts of "scratchpad register" and "one-time pads" from our intelligence community culture. Similarly the fast temporary "cache memories" and "cache registers" now embodied everywhere in computing systems large and small were originated by us in analogy with espionage techniques and terminology.

I needed four LGP-30's, linked together by punched paper tapes (!), to simulate the desired larger system. To manage that configuration (and please: do not let the moving tapes tear...), I needed to carefully program precise parallel computing control and timing of communications between the four machines. To obtain a position fix in the minimum time possible (about 30 seconds), I optimized the flows of the data and the mathematical calculations. Such methods later were called pipelining in the computer culture. My linked four LGP-30s may have been the world's first parallel computer.

I remember driving alone nights down to the Navy top secret net-control center at Cheltenham, Maryland, to install my four-computer LGP-30 parallel computing system in 1960 into 1961. Some Naval personnel were assigned to assist me. My prototype system would triangulate the Soviet submarines' positions and would provide 95 percent confidence search regions for them, and pass these on to NSA analysts and Naval commands. I began to wonder if my Army Basic Training three-month obligation was being delayed until I had my system up and running operationally. Therefore I decided to train some talented Navy chief petty officers to run my system. Off I went to Fort Knox to suffer the purposeful sleep deprivation and humiliation of Army Basic Training. The contrast with my life as a Navy computer intelligence expert was almost comical. Essentially I was treated as an intelligence officer in the Navy and became accustomed to salutes and saluting as a habit. The Army did make me a Corporal and squad leader, however. Finally, Becky and I could get married in June, 1961.

* * *

In the Countermeasures Branch, there was also a top-secret small satellite program under development by a small team headed by Reid Mayo. The historical origin of that potential satellite reconnaissance effort was actually the Moon. One of Lorenzen's long-time colleagues, Jim Trexler, had found that many Soviet secret electronic and communications signals could be obtained just from the reflections from the Moon back to Earth. A huge, 66-story tall moveable telescope was planned in West Virginia to collect those signals. So the Moon could be considered the world's first (natural) spy satellite! But for technical reasons only a smaller telescope was built. Nonetheless, the so-called Moon-bounce intercept capabilities stimulated Reid Mayo in 1958 to think about using man-made satellites to spy over Russia. Hundreds of pilots and technicians were losing their lives in ferret flights dodging about 200 miles in and out of Soviet border zones while trying to determine the electronic signatures of Soviet air-defense systems. Moreover, the satellite's intercept regions would be much larger than those of the ferret airplanes, and virtually unrestricted in orbits over the Soviet Union.

The United States had been relying on the U-2 spy planes to over-fly the Soviet Union until May 1, 1960, when Gary Francis Powers' U-2 was shot down.

This event had huge international repercussions and poisoned the relationship between President Dwight Eisenhower and the Soviet Union's Nikita Khrushev. Nonetheless, the need for electronic intelligence data was so paramount that only five days after the U-2 went down, Eisenhower gave approval for Mayo's group at the Naval Research Laboratory to send the world's first reconnaissance satellite into orbit over the Soviet Union.

Shortly after I returned to NRL in 1960, I saw Reid Mayo and his associate Vincent Rose laboriously plotting circles on a map of the Soviet Union. Although I wasn't cleared for their satellite project, as I watched I had a vague idea of what they envisaged. Just chatting, I told Reid that maybe I could give them their needed trajectories and intercept circles by writing a code to do that on my little LGP-30 computer. Reid jumped at the idea and we huddled for an hour or two to discuss exactly what was needed. Reid was of the older "fly by the seat of your pants" engineering mentality, and I could tell that he was skeptical that I could achieve the desired result.

In truth, I had my doubts as well. The LGP-30 had only 4,096 memory locations to play with. But I knew that machine intimately. And I already had written code which was doing geometry and navigational spherical trigonometry on the surface of the Earth for my first project, the submarine direction finding and tracking. My boss, Bruce Wald, was away for six weeks inspecting our installations at the receiving stations encircling the Soviet Union. I was alone in Washington and could work nonstop day and night to try to implement my idea.

In a furious effort that as I recall took me less than two weeks, I wrote new code which utilized as much as possible of the subroutines I already had written for the first project, and I succeeded. I still remember vividly the amazed look on the faces of Reid and Vince as I called them in to do a sample run for them. Their astonishment turned to rapture as we then ran some of their needed trajectories, and out of my computer came the desired coordinates and intercept regions.

This became my second successful NRL project. And until now, my having done so has never been recorded anywhere.

The world's first spy satellite launched into orbit on June 22, 1960, and carried the cover name GRAB (Galactic Radiation and Background) and code name TATTLETALE. Its principal purpose was to gather radar data, and any other electronic transmissions it might record. The computer codes I wrote were used for both tracking the satellite's position as it moved along its trajectory, and determining the intercept coverage area for each point on its trajectory.

It should be remembered that in those days, one could not even conceive of satellites with onboard computers. All collected intelligence data was processed afterward. The GRAB satellites only had antennas that would collect each pulse of a Soviet radar within a pre-specified bandwidth, and transpond a corresponding signal to NRL receiving stations at ground sites within GRAB's field of view. That intelligence data was collected on magnetic tapes which were then dispatched back to NRL and on to NSA. So much useful data was collected on the GRAB I satellite that the intelligence analysts at NSA were kept busy until GRAB II went up about a year later on June 29, 1961. That satellite then collected intercept data for more than 14 months.

As quoted in James Bamford's book about NSA (*Body of Secrets*, 2001, p. 366), Reid Mayo said, "The circle that we were able to intercept from instantaneously was about three thousand or thirty-two hundred miles in diameter, depending on the altitude." The consequences of the GRAB I and II satellites were enormous. The electronic intelligence collected by them was processed by NSA and marked a turning point in U.S. Cold War strategic doctrine. The Soviet air defense system was found to be too robust for penetration by SAC's (Strategic Air Command) high and medium altitude bombers, so these were succeeded by low altitude bombers and ballistic missiles.

I will never know how NRL and NSA partitioned the workload on the processing of all the electronic intelligence captured by the GRAB satellites. I was not cleared to know. But I do remember Reid and Vince begging me to train them so that they could operate my software on the LGP-30 computer. Then I remember them putting in long hours over many months hunched over the computer, getting out their needed intercept regions. Their team received the Presidential Service Award in 1962 for the world's first electronic intelligence satellite. I was a key part of that team, although not officially because I had never been cleared for it!

I returned to my primary tasking, that of position fixing from intercepted signals. That project, as we expanded it to industrial scale, too was successful, and became known as the Bullseye project. I wrote six reports on it which are still classified and unobtainable, even to me. I never wrote any report on my high intensity but unofficial work on the GRAB project.

The Air Force had opted for a photo-reconnaissance spy satellite program, the Corona satellite. The first successful one was launched in August 1960. Evidently the U.S. government was not satisfied with the Air Force program, and created the top-secret National Reconnaissance Office (NRO) in August 1960 to take over supervision of the Air Force's spy satellite programs. When NRL failed in its GRAB III satellite in 1962, NRO also took over supervision of NRL's ELINT satellite activities in July, 1962. I was never told of NRO's existence. The satellite intercept data my computer codes help collect was, in my mind, always going to NSA or CIA.

Bamford's book (2001) is an excellent source for further dramatic instances of national significance in which my computer algorithms or their successors (on better computers) served the nation. He gives a compelling account of the Cheltenham Naval Station's role in tracking Russian submarines during the almost-nuclear war situation which arose in the 1962 Cuban Missile Crisis. There, my four-computer LGP-30 system could quickly return the Russian submarines positions, and as well, the positions of virtually all of the Russian surface ships, every time they communicated among themselves or with Moscow, at least ten times faster than the previous manual plot-by-hand Naval system. Time was truly of the essence in that nerve-wracking nuclear confrontation.

Also in Bamford's book you will find accounts of important unforeseen consequences of Mayo's GRAB program. For example, a treasure trove of electronic signatures from the Russian space program was inadvertently collected when the Russians lost critical contact with a cosmonaut and "turned on everything" to try to re-establish contact. Additionally, the first Soviet antiballistic missile radar signatures and the locations from which they emanated were obtained by GRAB.

* * *

This account of my life's trajectory placing me as "the right man at the right place at the right time" would have been lost from history. Only by an accident of fate is it now written.

When I left NRL in 1963, turning my attention to full-time graduate work in mathematics at the University of Maryland, my extremely brief debriefing instructions were curt and clear: "Never talk about this work." I never did. And I didn't happen to read Bamford's book, published almost 40 years after my work, on direction finding and satellite spying. Thus I had no idea that the GRAB program had been partially declassified in 1998 in order to reveal its important contribution at NRL's 75th anniversary celebration, and further partially declassified in 2006 to allow an exhibit of those early Navy spy satellites at the National Cryptologic Museum.

On a whim, I decided to go to my 50th Class Reunion here in Boulder, for the University of Colorado Class of 1958. I paid my $125 to the Alumni Association and prepared for two days of being wined and dined by my own administrators here. Little would some of them know that they were supplicating themselves to one of their own employees! But the big surprise would be to me. As a result, I would decide to write this book.

The very first morning when I went to pick up our caps and gown to wear in the Commencement procession the next morning, within 10 minutes in walked Gordon Fink, a fraternity brother I had not seen for 50 years.

"Hi Gordon," I greeted him, as if only a semester had passed since our last meeting.

"Hi Karl," he replied, without skipping a beat.

We chatted. Gordon had risen to the very highest levels of NSA, CIA, DEA, even NRO. Somewhat in awe, I said, "Gordon, I am going to tell you something that I have never told anyone. I wrote the software for the world's first spy satellite in 1960. There at NRL. Right after Gary Francis Power's U2 was shot down over Russia."

"Karl, I have seen that satellite," Gordon replied. "The program was recently declassified. Do you want me to send you a picture of it?"

Had that chance conversation with Gordon not taken place in 2008, I might never have written this account of my role in history. For that matter, I might never have written this tale of my life. Chance is like that, isn't it? We never know what it will bring.

* * *

Large-scale direction finding networks emerged in World War II as a countermeasure against German U-boats, which were sinking large numbers of Allied vessels in the War of the Atlantic. Our later computerized systems were still of course employed to locate adversary submarines, but were also invaluable in their use for general electronic surveillance of the Soviet Union and other potential adversaries. Now such systems can fix the locations of terrorists just from a cell-phone emission. Such devices use line of sight propagation paths. If you have antennas of sufficiently high gain, you can detect even the weakest of signals. Transmissions from far space planetary exploration satellites are a good example. Because electronic transmissions generally go out in spherical waves, in principle they can be detected from any line of sight direction.

The Soviet launch of Sputnik and its over-fly of the United States had the unintended consequence of opening their country to satellite reconnaissance. President Eisenhower soon ordered full speed ahead on United States satellite systems to perform such electronic espionage from space. GRAB just happened to be the first spy satellite. Within the years 1959 to 1963, the world of espionage went into space.

Espionage prevents wars and thereby saves millions of lives. I like the prize-fighter analogy again. If you and your opponent can each detect the buildup by the other of potential kayo punches before they are thrown, you are both going to end up spending your 15 rounds just feinting and jabbing as you maneuver around the ring. There are no surprises and no knockouts.

I firmly believe that the advent of satellites, although originally motivated by the need to gather electronic intelligence, fundamentally and forever changed the dynamics of our human society on Earth. It took us from life in two dimensions to life in the three dimensions of space. It isn't possible now, 50 years later, to even imagine living in the pre-Space age. Human society and its promise and perils have been utterly transformed.

The GPS (Global Positioning Satellite) system now so common in our lives can serve as an example of this transformation. GPS may be thought of as direction finding in reverse. Instead of silent receiving stations waiting to triangulate a bleep from a submarine, the GPS satellites are always bleeping and you want to triangulate off them. We are all submarines now, under the great sea of space.

Notes

I have given a more detailed unclassified account of my specific mathematical and pioneer computing contributions to the primary task of tracking Russian submarines in my paper *Parallel Computing Forty Years Ago*, Mathematics and Computers in Simulation 51 (1999).

You can find more information about the 1960 satellite work and consequent further developments from a number of sources about the GRAB program, now available on the internet via Google. Also in Bamford's book, *Body of Secrets*, Random House (2001).

Some further information can be obtained from NRL Press Release 32-00r on 5/22/2000: *A Tribute to the Father of Electronic Warfare*. This was an account of the crucial role of our Branch Head, Howard Lorenzen, who died 2/23/2000, throughout his life of military electronic intelligence work.

5. First Publication

...and a very strange letter...

I had kept my promise to the military. Not only had I fulfilled the primary task of my critical skilled enlistment – computerizing direction finding – but I had also by chance significantly advanced the GRAB intelligence satellite program. When I asked my Navy superiors in 1962 if I might reduce my commitment to 20 hours a week, in order to begin graduate school in mathematics at the University of Maryland, they were happy to oblige. The only condition was that I would be "on call" if they needed me for anything urgent.

I took a Teaching Assistant position within the Mathematics Department to better assimilate into the grad-student culture on my way to a Ph.D. But after one year of juggling the NRL job, the TA job, and carrying a full graduate course load, I was exhausted. So in 1963 I asked my friends at NRL if I could officially resign, with the unofficial promise that they could call on me anytime they might need me. Again we were in happy accord.

I found myself already a little older than most of the mathematics graduate students. And when my daughter Amy was born in April, 1963, I had fatherhood and supporting a family to add to my responsibilities. The mathematics program was no piece of cake either: One needed to pass three required graduate exams before going on to Ph.D. research work. These were Analysis, Algebra, and a third, which was a potpourri of other mathematical topics. All summer in 1963 I studied algebra books and worked algebra problems, having never taken a graduate-level algebra course. I was gambling. But much was at stake; I had to succeed. I passed the Analysis exam, scraped by on the potpourri exam (using what I'd learned in courses on differential equations and algebraic topology), and scored highest on the Algebra exam.

I decided to skip the Master's degree and head straight for a Ph.D. I abandoned algebraic topology as too obscure for my tastes. I took a number of courses in probability and statistics and wanted to delve into the fascinating dichotomy that exists between those two subjects. But in a stroke of good luck I was offered a Research Assistantship in partial differential equations, and worked on a Ph.D. dissertation under Professor Lawrence Payne. Larry was a terrific adviser and gave me a choice among several potential problems. I chose one on the modeling of nuclear reactors and wrote an 88 page Ph.D. thesis entitled *A priori bounds with applications to integrodifferential boundary problems*, a rather technical work without any dramatic conclusions. I received my Ph.D. in June, 1965.

The 1960s was a golden decade in which to obtain a Ph.D. in mathematics. I could go into teaching at a university. With my computing experience I could enter into a career with IBM or one of its competitors. I could even go back to the projects in electronic espionage at NRL. Of course always in my mind was a wish to immediately get back to Boulder. None of those outcomes transpired!

A fellow graduate student and friend, J. Keith Oddson, urged me to apply for an NSF-NATO post-doctoral fellowship in Europe. That was in retrospect very generous of Keith, because we would be competitors for those awards. Keith was going to apply to work with an Italian professor, Carlo Pucci, in Genova, Italy. I asked Larry and he said, sure, apply to work with Norman Bazley in Geneva and Gaetano Fichera in Rome. My wife was all for it – who wants to be a corporate wife? So I wrote as good a scientific proposal as I could. I really did not give it much beyond a hope. But both Keith and I were selected. I heard afterward that there had been only four awards out of 27 applications. So off to Switzerland we went.

During my six months, July to December of 1965, there in Norman Bazley's small Advanced Studies Center at the Battelle Institute in Carouge, a suburb of Geneva, I had complete freedom mathematically. One day I came across a 1964 paper by Ed Nelson called *Feynman Integrals and the Schrödinger Equation*. In that paper Nelson was able to prove a certain perturbation theory important for quantum mechanics. Nelson employed a perturbation B that was restricted to one-half the size of A. I saw how I could, using a sequence of inequalities, extend B all the way up to any size less than that of A. Accordingly I wrote a short paper of five pages entitled *A Perturbation Lemma*.

I asked Norman his opinion of my paper and he was very positive. He said, "Send it to Browder." Felix Browder was one of the most important American mathematicians and one of the editors of the prestigious Bulletin of the American Mathematical Society. So off I sent the paper. I was a complete unknown to Browder. But he published the paper, and it appeared in March, 1966. It was my first mathematical publication.

Little did I know that a number of top mathematicians had been trying to get my result. I just thought I had been lucky, and that Browder had been in a good mood that day. It is only now, as I look back at my life, that I can see how that paper opened the door for me into academia. Offers from Berkeley, Stanford, Harvard, and Minnesota would be forthcoming. I would eventually choose Minnesota at Larry's urgings, as it was the best school in the field of partial differential equations.

We went down to Rome for the second six months of my post-doctoral work. There I would present my first mathematics colloquium talk. Naturally, I presented my new result. There followed a lively debate about its merits between Gaetano Fichera and his Russian visitor, Olga Oleinik, a very famous (and jolly) specialist on partial differential equations. Afterward we all went out to dinner. There Terry Zachmanoglu, a Fulbright visitor to Fichera's institute, took me aside and said to the newcomer (me): "Don't worry about anything they said about your result. They were just trying to impress each other!"

The summer of 1966 was spent at the Forschungsinstitut in Zurich. My officemate was the famous topologist John Milnor. We picnicked on weekends with Cleve Moler and his wife and children. Cleve went on to form the important mathematical software company MATLAB.

Fall took us to the University of Minnesota where I began teaching as a tenure-track Assistant Professor. But soon we were headed back to Switzerland. I had accepted a second postdoctoral at Norman's ASC in Geneva for January through August, 1967. Upon our return to Geneva, Norman handed me some mail that had come during my absence. I dutifully opened each letter. Then, in an envelope with no return address, I came upon a very strange letter.

Handwrittten, dated Nov. 17, 1966, three words were scrawled at the top of the page in heavy black ink:

Persuasion and supplication

The letter, addressed to me at the Geneva address, Battelle Institute, Genf, Schweiz, continued in blue ink as follows:

"Dear Mr. Gustafson,

As a symbol of mathematicians and their relation to world movements, etc. let me make a petition to U concerning time.

I'm a slave, in effect, and could be viewed as a slave....but how long, how long, should this slavery last?"

The letter continues, and in the next paragraph states:

"Then there arises a Chinese symbolic theory concerning the time. It seems the Genf parliament has 100 members. 1,865 + 100 = 1,965. 1,815 + 150 = 1,965 (150 ~ communication) communication would suffice to free me, and would be a good liberator if of the right sort (safe)."

The letter goes on in this fashion: The numbers 101 and 151 are said to have 40 primitive roots. 40 links with the ideas of Exodus and with Moses, Mohammed, Esau. Significant events occurred in the year 1926

"...therefore it seems to me plausible that U of the left, as it were, could think 1966 a good time to let slavery end."

The next paragraph introduces (in green ink) the quintal polynomial:

$$v^5 - 13v^4 + 65v^3 - 155v^2 + 174v - 72$$

as representing five human beings who may be thought of as born simultaneously with liberation. The years 1926 and 1966 are said to be related by field-particle duality. The next paragraph identifies the writer as 38 years old. The description of John 5,1,15 of the pond of Probatica and the man able to take up his bed and walk remind the writer of the Vierwaldstattersee and New York City.

"...And David was the grandson of Obed."

Then the writer turns to Job theory, with the equations

$$1,828 + 140 = 1,968 = 1,928 + 40$$

and

$$1,828 + 137 = 1,965, 1,828 + 138 = 1,966$$

and wherein 1965 is related to Vietnam, 138 to Joan of Arc (symbolically), and $137 = 411/3$, to the coarse structure constant [he means: the fine structure constant of quantum mechanics, which is only approximately 137].

At the bottom of the second page, the letter closes, "So let me beg, beg, beg, beg at U that this wasting time be shortened, be reduced as much as possible!

Hopefully,

John Forbes Nash, Jr."

I was so dumbfounded as I read the letter that even before I reached the end I was walking down the hall to Norman's office. I had not even read the signature, but when Norman got to it, he said, "My God, you've got a letter from John Nash, America's greatest (mathematical) analyst!" I stood unblinking for a moment and then asked Norman, in bewilderment: "What do I do with it?" He replied matter-of-factly, "You should save it." So I put the inscrutable letter in my N-letter folder. When John Nash received the Nobel Prize in Economics in 1994 for his mathematics of game theory, I looked in my N-folder. There was the letter.

Of course from my knowledge of partial differential equations I knew in 1966 of the great Nash-de Georgi estimates for elliptic partial differential equations. I also knew vaguely of the Nash Implicit Function Theorem. But why would Nash write to me? Norman said Nash sometimes spent time in Geneva or Paris, and suffered from mental breakdowns. I let it go at that.

In retrospect, my only explanation is that Browder, who was a friend of Nash, might have shown my first paper to Nash when I submitted it to the Bulletin, or that Nash might have just happened to see it when it appeared. I cannot see any other way than through my paper that Nash could have known my rather remote Battelle address in Geneva. But his letter made no reference to either my paper, or why he chose to write to me. It was to remain mostly a mystery.

The 1998 biography *A Beautiful Mind* by Sylvia Nasar gives an intimate account of John Nash's life and fight against paranoid schizophrenia. That book was made into an Academy Award winning movie of the same title. When I decided to write this autobiographical account of my life, I emailed Sylvia Nasar at Columbia University, where she is a professor of business journalism, about my 1966 letter from John Nash, and how she thought I should handle it. She replied the same day, saying to just make a copy of the letter and mail it, along with my request to Nash, in care of the Princeton math department. That I did, on May 21, 2008. In my cover letter I mentioned that I might want to include the letter in some way in a book I would write, and would he grant his permission?

On May 28, 2008 I received a fine, lucid response from John, via email, which began: "That was one of my 'mad letters.'" He went on to explain that by the summer of 1966 – after having spent many months in a mental hospital in 1965, and experiencing a subsequent short-lived return to rational thought – he had again fallen into a "delusional status." That status persisted until the 1980s, he said. He could not remember precisely what had led him to write to me. But he recalled that he had special thoughts about Switzerland during his "mad times," so he surmised that the Battelle Institute address was likely what motivated him.

Nash's unusual letter carries fragments of facts and fantasy. Recently while perusing it, a pattern suddenly appeared to me, after all these years. Not being a number theorist, I decided to check if indeed 101 and 151 each has 40 primitive roots. And indeed they each do. Hmmm...the number 40, what was the significance there, I wondered? Nash wants "to let slavery end" in 1966. Hmmm. When he wrote the letter, the last quadrennial International Congress of Mathematicians had already concluded. Nash knew then that he would never win the Fields medal, being 38 years old and having been passed over in 1958 (his best chance), in 1962, and finally in 1966. The Fields Medal is considered the top honor a mathematician can receive. Nash would not get another chance, as the coveted award has an age limit of 40, and the Congress meets only every 4 years. "I am 38 years old," he writes in the letter. Then comes the calculation 1928 (Nash's birth year) + 40. Wow. In a flash, it dawned on me that it was all about Nash's realization in 1966: He would never be a Fields Medalist, the goal he had pursued with such fervor for so many years.

Although John Nash likely had some genetic predisposition to schizophrenia, I believe that many mathematicians can focus so intently on a set of conditions characterizing a mathematical problem that these conditions can take over the mind, to the exclusion of conditions essential for a healthy mind. I am much more like John Nash than I ever realized early in my career. In math, intuition can leap too far ahead of reason. Even more, one can experience states of extreme bliss while engrossed in a mathematical problem, and this can be quite overpowering, like a drug pulling you in, more and more, deeper and deeper, closer and closer. And then you are gone. Completely gone, and the mind is gone. You begin to cope with reality only in incomplete fragments at that point. But you cannot come back...and why come back? Believe me, it is fantastic to be off in that other world, exploring your own new mathematical territory, so why pull out?

Nash alluded to this experience in a short autobiography he wrote for the Nobel Foundation in 1994. (Later published in *Les Prixes Nobel,* [Nobel Foundation], Stockholm, 1995.) Reflecting on his return to rational thought, he lamented that "this is not entirely a matter of joy as if someone returned from physical disability to good physical health. One aspect of this is that rationality of thought imposes a limit on a person's concept of his relation to the cosmos." And could it be that, to some extent, this limit may also dim brilliance?

I occasionally use Nash's letter as a point of illustration in my course lectures. Recently, I used the letter to motivate a lecture on Game Theory in a course on world population explosion. As noted above, Nash received a Nobel Prize in

Economics for his mathematical formulations of Noncooperative Games. I view the world population explosion as one of the most important, even critical, noncooperative games of our time. I argue that most environmental movements are just Bandaids by comparison. It never ceases to annoy me that Al Gore (with his four children) and the climate change scientists and politicians don't even mention the overriding and pressing need for more rigorous worldwide population control.

* * *

My postdoctoral years of 1965–1967 exerted a powerful influence on my thinking. My mentality had been basically that of an engineer, or at best that of a scientist, with emphasis on the practical and not on the theoretical. Those years instilled in me more appreciation of the theoretical, especially toward physics. Since then I have always been half European, half American in my view of what a professor should be. In Europe, if you meet someone for the first time, they ask you, "What is your position?" In America, it is, "What do you do?" In Europe, there is genuine respect for intellectuals. In America, there is often a certain disrespect, or a suspicion that maybe you aren't really earning your salary! Believe you me, I earn my salary, probably working 3,000 hours some years. But some of my colleagues who have dropped out of research put in less than 1,500 hours per year. Top professors must stay active in academic research. That is easier said than done.

The job has a wonderful feature of some time flexibility. I think that is essential for mental health. One cannot push the mind to its limit without some respite and down time. Sometimes when I push hard for a month on a single mathematics problem, thereby also subliminally working on it at night while trying to get some sleep, I will start having strange and even bad dreams. Through awareness of my tendencies, I have come to learn to just force myself to pick up all the pieces of paper I have written on a problem, stuff them into a manila file, put it on the shelf, and turn my mind to other things for a while.

I had minimal preparation in foreign languages before my postdoctoral years of 1965–1967 in Europe: One year of high school Spanish, and two cram courses in French and German at Maryland which were specially designed to get Ph.D. candidates past their required foreign language requirements. As an Engineering undergraduate, I took no language courses. It was quite a shock to wake up that first morning in Geneva, Switzerland, unable to say a word to the proprietor at the small hotel where Norman had put us. My wife Becky however had taken several years of French in college, so she could communicate for both of us. Her adeptness in the language also helped her to enjoy our new life in Swiss Romande, the French speaking part of Switzerland. Of course everyone at the Battelle Institute spoke English. Our daughter Amy was only two to four years old in this postdoctoral period, but when we returned to Switzerland for a year in 1971, we put her in public school and she became fluent in French in about two months. I am sure that all those bilingual sounds in her childhood were of benefit to her. In some ways, she is part European in mentality, like me.

All told I have spent almost eight years, or about 10 percent of my life, outside of the United States. Substantial parts of that were in Geneva or Lausanne in Switzerland or in Brussels, Belgium. I do not have a good ear for languages, or for anything else, for that matter! I seem to be extremely visual. I don't know how I survived school and college, although the answer is evident: visual retention. I can listen to a philosophy professor chat away while perched on his desk for an hour, I understand and follow all of his reasonings, and fifteen minutes after class, it is all gone. I "hit the wall" for the first time in my junior year in university and learned to start jotting down on paper little notes about what I was hearing. Often that was sufficient to allow me to reconstruct the main gist of what was said.

It was inevitable that gradually I started learning to speak French in Geneva in those postdoctoral years. Then in 1971–1972 on my sabbatical year back in Switzerland, I gave a lot of lectures at the EPFL in Lausanne (École Polytechnique Fédérale de Lausanne) and became more comfortable in speaking French. When I was invited to give a three-month advanced mathematics course there in the summer of 1975, I decided to present the course entirely in French. To do so, I totally immersed myself in the local culture of that lovely city by Lac Leman, always eating out, some evenings in bars, weekends in mountain villages, the like. It worked and I became reasonably fluent and certainly comfortable in speaking French. Moreover I seem to have almost no accent. But it is street French. I still cannot write a good letter *en français*, and my grammar is automatic without any knowledge of what tenses I am really speaking.

Mountain climbing and skiing are of course incumbent upon any climber or skier visiting Switzerland. In our years there of 1965, 1967, and 1971–1972, my family and I greatly enjoyed ski holidays at Wengen, Murren, St. Moritz, and Verbier. Also I could ski day trips at small ski stations near Geneva. One of my favorites was La Clusaz. When the weather was good, the pistes were great, reminding me of Aspen – but they were better in La Clusaz. When the weather was bad, because the ticket system was just that of buying individual rides up the lifts, you bagged it and went into the restaurant and had a croûte, a wonderful Swiss fried creation of egg on top of sizzling cheese over some French bread. Over the new year's holiday in 1971–1972, we spent two weeks sharing a chalet with friends at the newer Swiss ski resort Verbier. My daughter would go out the door, step into her skis, and downhill to ski-school. We all had such good times over there.

One day while in Wengen, my British physicist friend Michael Boon suggested that he and I take the train up to Kleine Scheidegg under the Eigerwand and ski down to Grindelwald, then have lunch, train back up to Kleine Scheidegg, and ski back down to Wengen. Morning's nice weather deteriorated into a full-scale blizzard in the afternoon as the train neared the pass at Kleine Scheidegg. The train stopped short of the top and the conductor announced, "Eigergletscher!" This was a very little piste served by a Poma lift right under the Eiger's north face. Michael looked at me and we grabbed our skis and jumped out. There were five of us standing there, four Brits and me. One of the Brits exclaimed, "My God, only mad dogs and Englishmen would get out here!" I chipped in, in my American accent, shouting against the wind, "Hey! I resent that statement!" We all skied that steep little run several times in near white-out conditions, in what I might say was definitely mad-dog ferocity.

It was in these postdoctoral years of 1965–1967 that quantum mechanics really cast its spell on me. Although I had to choose at Maryland whether to attempt my Ph.D. in Physics or Mathematics, I had chosen mathematics because I'd already taken a year's graduate course in pure mathematics at Colorado. I also felt that I might get lost in a big Physics Department, and that my chances of success were higher in mathematics. So my knowledge of physics was all at the undergraduate level. However, Norman Bazley had set up a joint seminar with Professor Josef Jauch at the Institut de Physique Théorique of the University of Geneva. Jauch's main interest was quantum mechanics and he wrote important books on the topic. By chance I was an expert on operator theory, the mathematical language in which the theory of quantum mechanics is formulated. For example, Schrodinger's partial differential equations, and Heisenberg's matrix mechanics, are linear operators.

Quantum mechanics, as everyone knows, is a really strange business. Einstein once complained that he worked two hundred times as hard on quantum mechanics as on his theory of relativity, and never understood the former. So much of how we use quantum mechanics is based upon its mathematics without being able to really understand the details of the underlying physics. An example is Schrödinger's partial differential equation, or Heisenberg's matrix mechanics. But to really do physics, you must develop physical intuition. I was a late starter to quantum mechanics, and I am still after 40 years as frustrated as Einstein was. I think I have developed some physical pictures of what might be really going on, but I will probably never know for sure. To know for sure, we need to physically, by delicate physical experiment that's so far impossible to carry out, be able to better probe the true nature of the interior of an electron. I may be attacked for such a statement; physicists love to attack each other. But that is my current intuition.

Mountain climbing beckoned in Geneva and a rather unsafe manager at Battelle, a Monsieur Paquet, at first took me up the Saleve, a cliff above Geneva. I noticed his use of rope and belays was terribly careless. I felt relieved when we topped out. Then he suggested the Aigle du Moine, above Chamonix. We did that, again unsafely. Coming down, the rope stuck on a rappel and M. Paquet blithely just pulled himself up it to free it where it was stuck. Later on I skied the Mer de Glace, the large glacier coming off the north side of Mt. Blanc, with Larry Horowitz, a physicist. Michael Boon and I climbed up to the Aiguille de Gouter hut at 12,500 feet to attempt Mt. Blanc, but after two days of blizzard in which no one could go out of the refuge, we came back down. Although that is the tourist route up Mt. Blanc, you must cross a dangerous rock-fall couloir on the way up to the hut. The first time we crossed it, we had to wait for a party of Italian climbers, each of whom said his Hail Marys, then sprinted across in his crampons. Our turn came and Michael went, then me. As I took off my crampons, a small rock invisibly whizzed by my head, missing me by about a foot. Rock fall is a real hazard in the Alps. Later in the summer we tried again. This time we made the summit on a beautiful clear-sky August day. We took our full packs to the top, leaving nothing at the hut, thinking we might prefer to just descend directly via the large icefall so clearly massive and visible from Chamonix, rather than cross that infernal couloir on the ascent route again. This was actually the first large glacial icefall either Michael or I had negotiated, and we had no guide. Due to the summer heat, only an almost imperceptible track, which soon disappeared, led down. Nonetheless we strode downward and I found it great fun to thread one's way between seracs and around crevasses, walking on your crampons almost as if on concrete in some wonderland made by Disney. Late summer had exposed the crevasses, although we met problematic snow-bridges, crossing some, doing long traverses around others. At about 8,000 feet of descent below Mt. Blanc's summit, you arrive at a trail that traverses you over to a midstation of the tourist telecabin that runs up and down from Chamonix to Aigle du Midi. Congratulating ourselves on avoiding the rock fall danger of the couloir on the other side, we were surprised on this traverse by a big rock fall of very large blocks of rock that came roaring down from on high and scattered across the trail just in front of us. But it missed us. A truly glorious day.

* * *

My host Norman Bazley at the Advanced Studies Center at Battelle in Geneva was a most unconventional, even controversial, character. He had obtained his Ph.D. at Maryland by solving a very challenging mathematical problem in quantum mechanics, that of providing lower bounds for the eigenvalues of the chemical elements. Norman joked to me, "Karl, you know, I wrecked my car at an intersection working on that problem!" Totally in rebellion against the usual norms of professional demeanor, Norman led a life of a lot of drinking, dining, and partying in general. If you believed what he told you, he was a confirmed womanizer.

I will never forget a trip with Norman for a week's research in Rome in connection with the group of Professor Fichera there. We were three from Battelle, Norman, me, and a Swiss postdoctoral at Battelle, Bruno Zwahlen. Most of the time, don't we recount our day's events beginning in the morning? With our days with Norman, I have to start in the evenings. After a great Italian dinner, the wine kept flowing, along with some stronger liquors. Then one goes to late night clubs. Always drinking and always talking to some extent about some mathematics. Finally at 3 a.m. you find your hotel. The next day you wake up late with a bad hangover. The group assembles to sober up for lunch and does some math before heading out for dinner. I personally do not feel like having any more after several drinks, and usually stop. With Norman it was the opposite.

The hard living had caught up to Norman already and he was on four different pills: one for his heart, one for blood pressure, one for nerves, and one for gout. These did not change his behavior although he did have to watch his diet. With his friend and colleague David Fox, who was also a Maryland Ph.D., Norman formed a computer company and sold stock to his Battelle colleagues. This is a cardinal sin in a startup company's financing and sure enough the company went bankrupt, angering his Battelle superiors. At the tipping point at Battelle, suddenly in 1972 Norman was offered a full professorship in Köln, Germany. This was in the era of the Baader-Meinhof left wing guerilla group's violent activities in West Germany. When I later gave a lecture in Köln in June, 1976, I accompanied Norman and a girlfriend and her pet poodle late at night to a bar that he boasted was frequented by the guerillas.

Norman always seemed to have enormous travel grant support at Battelle via the U.S. Army command in Frankfurt. That was highly unusual. In mathematics, no one gets a blank check like he did. Then during the 1973 Allende crises in Chile, suddenly Norman had a visiting position there. When I gave a lecture in 1980 in Köln, he introduced me to an attractive young female assistant professor from Chile who was in residence in Köln. Norman told me she was very hot. Another story was of his close call with his Jamaican girlfriend in New York. The bar they were at was robbed, all patrons asked to lie face down on the floor, shotguns to their heads, but only their billfolds were taken. On his way to Australia in 1981, he seemed to have to stop over in Tonga where he found a girlfriend to spend a week in a straw hut during a time of turmoil there. Norman's behavior was always that of a boisterous swinger and partier, very open, life of the party.

Although married to a loyal second wife, a nice German lady we all knew and liked, Norman seemed to have a favorite mistress in Köln. Elizabeth was a pharmacist, lithe and sinewy, and sometimes accompanied him on his travels. I was consulting in Switzerland in 1989, shortly before the reunification of the two Germanys, and Norman and Elizabeth came down to our Swiss friend Pierre's house for a dinner party. After an appropriate number of glasses of wine, in front of Elizabeth, in his open do-you-believe-it style, Norman complained in high voice, "The East Germans have published a letter in their newspapers from my girlfriend over there, and she claims I am a spy for the C.I.A.! Can you believe that?" looking at Pierre and me. Pierre and I looked at each other, then at Norman, and both said, well, yes Norman, we could believe it. Elizabeth said nothing, just maintained her dutiful pose. Norman continued, "Well, I'm ruined over there; I can't go back to my visiting position there..." and continued to grouse.

That was the last time I saw Norman. I was sad to hear of his untimely death a little more than a year later, and of how it happened. It seems he was at a party on a Friday afternoon in Köln and somehow fell off the balcony of the fourth-story apartment. Then he was recovering in a hospital but had a mysterious relapse and died.

Pierre's opinion when we later discussed Norman's strange life was that Norman may have been a recruiter, setting up young espionage contacts, especially female agents, at trouble spots as requested by U.S. intelligence services. Certainly his open partying lifestyle gave him perfect cover to do so. In support of this is the fact that his wife never openly complained about his obvious womanizing – of course, the cover for that was that she was a German hausfrau. We will never really know as those Cold War years fade into obscurity.

No matter whether Norman had such a second tasking, to me he was full of life, a lot of fun, always helpful, and always liked talking mathematics and physics.

* * *

There is no doubt that my postdoctoral years in Europe profoundly changed my life. A chance suggestion by a fellow graduate student had led to an internationalization of my perspective. I learned to love fine wine and sizzling cheeses. My mind became partially bilingual in thought pattern. I was exposed to great theoretical physics and committed myself to trying to understand some of it. The other postdoctoral students at Battelle with whom I had become acquainted went on to become professors in Switzerland, Germany, and France. Some of them tried to hire me away from Colorado to the EPFL in Lausanne during my 1971–1972 lectures there. But my wife Becky, who greatly enjoyed our sojourns in Europe, nonetheless vetoed the offer. She was right. I am an American. Switzerland, although beautiful, keeps you there as a foreigner. It is not like here where we readily integrate newcomers from all countries.

Notes

Our six months in Rome took me to the Institute of Mathematics at the University of Rome. There were no stop lights, just roundabouts, as I drove to work from our apartment at 36 via Monte della Gioie in the northern part of the city. Student riots, although not very damaging, were common as I passed through the gates of the university. My host Gaetano Fichera taught only when he felt like it, usually the two months from February to March each year. Because I knew the recent unbounded operator theory, he pressed me to write a book on partial differential equations with him, but I didn't have the time to do so. Based upon my small knowledge of Spanish, I became passable in speaking Italian while we were there.

The three months in Zurich summer 1966 was mainly a holding action before we returned to the United States to my position as Assistant Professor at the University of Minnesota. In Zurich, we were faced with a third foreign language: Schweizerdeutsch! Now, that is truly a tough one. Much later I would consult at ABB in the nearby town of Baden and actually learn a few words in the local lingo, beyond my small knowledge of usual German.

I met Pierre Furrer in 1981 on his very first day in the United States. The ever-gregarious Norman Bazley, who had met Pierre on a transatlantic flight that same day, brought him to a party at my house that evening. Pierre and I became lifelong friends. We had great times together whenever I visited Switzerland. But all the unrestrained eating and drinking may have contributed to his death at age 64 in 2010.

6. Into Academia

...despite the politics...

When my Ph.D. advisor Larry Payne heard I was leaving Minnesota to go to Colorado, he told me, "Sure, I understand, you want to go home...but you know, that Math Department out there has always messed things up. Almost always, they do." Larry seldom said anything bad about anybody. But his words would turn out to be prophetic indeed.

In my more than 40 years here since 1968, I have become a world-class and widely recognized mathematician. My 270 published papers put me well within the top 1 percent of all published mathematicians worldwide. No one in the history of Colorado mathematics – all universities, all departments, pure or applied – has come even close to that achievement. Additionally, I have published 15 books, given 300 talks and 100 international invited speaker addresses in 35 countries. No one currently in mathematics at the University of Colorado has anything near those numbers. Yet my achievements have been studiously ignored by my department, and my academic salary is substantially lower than the salaries of several of my colleagues.

There are several reasons for this at first seeming incongruity. One is Larry's prophecy. Another is my preference for breadth of expertise over narrowness of expertise. Probably the most essential factor is my lack of cunning. Beneath their liberal rhetoric and innocent posturing, I have found my pure mathematician colleagues to be clever conservatives, especially when an issue affects their own vital professional interests. Maybe such is not uncommon throughout academia. But pure mathematicians are a special breed. By nature and choice they exhibit an extreme narrowness of outlook. From it they develop what I have come to call "an absoluteness."

I now have a little of this absoluteness of focus in myself. I don't think I had it when I was growing up. I don't think I had it as I was being trained in engineering, physics, and business finance. I don't think I had it even when I was first an applied mathematician. But I had to get some of it to succeed in pure mathematics. You cannot compete in pure mathematics unless you can concentrate on a single problem over a long period. This means you sacrifice a lot, very often a lot of your social life, and of necessity you must focus on your own mental effort.

With such a large investment of your energy and time on a single domain in which you must compete worldwide with very intelligent others of the same mindset, you come to believe that "your domain" is the most important domain.

Many years ago we had a nice coffee room in the department where colleagues could gather to chat about the mathematical research problems they were grappling with. Over the years I noticed more gossip and less mathematics: Who in the department was doing good work, who was not. One day, in what seemed like yet another character-assassination session, my friend Larry Baggett – an excellent mathematician who has been blind since age 5 and, perhaps because of that, hears in sharper ways – sat back in his chair and chuckled. "Let's see if I have this right: If someone produces fewer papers than I do, he is not active. And if he publishes more papers than I do, then he is not deep." There was palpable discomfort in the room; but Larry's admonition was simply not welcome, and it was quickly forgotten.

It is well-known that university life has a lot of in-fighting. There is that joke about why Vice President Hubert Humphrey left Macalester College in St. Paul, Minnesota, to go back to the U.S. Senate. According to Humphrey, *the politics were not as nasty*. That there is relatively little money in higher education is certainly a factor creating tough battles. But within mathematics, the strife seems to take on an almost religious tone. It has reminded me of the ridiculous struggles of my father and mother as they fought over whether to send my brother and me to Lutheran or Baptist Sunday School. We really didn't care. So it must come down to one thing: Power. Enough power to ensure your academic survival.

I have decided not to place onto this book the incredible burden of describing the details of the nasty politics I have seen, and suffered from, here at the University of Colorado. Indeed, I could write a whole book about it. Would such an account change anything? I doubt it. Nonetheless, I have prospered mathematically and scientifically despite the politics. That I could do so no doubt depended on the quite different mental preparations and perspectives I had attained before returning to Colorado. I had gained a much, much broader experience and competence scientifically. The offsetting youthful character-building period of descending from upper middle class to middle lower class also helped. There was already an underlying philosophical tendency to want to see the whole picture about anything. And my postdoctoral years and experience in Europe gave me a vision of what makes a world-class professor, as well as a growing confidence in my own ability to be a strong professor.

So I will just give you an account of some of the events that transpired here, and how they required some course-changes in my life. One thing was clear: Now that I was back, nobody was going to chase me away from Boulder.

To switch on some better humor, let me mention that the advertisements and posters for the conferences on the Foundations of Quantum Mechanics held over the last ten years in Sweden, to which I have been one of the invited speakers four times, list me as Karl Gustafson, University of Boulder. Thank you! Thank you! Thank you!

So on with the story. After one winter in Minneapolis at the University of Minnesota, my always-held desire to get back to Boulder gave me no hesitation at all in accepting Stan Ulam's offer in 1968 of a tenure-track position in the Mathematics Department of the University of Colorado in Boulder. That year Ulam hired three mathematical physicists into the Mathematics Department to attempt to build a joint Ph.D. program with the Physics Department. That we did, and it ran well for a number of years. It then faded as interest in theoretical physics declined nationwide, in favor of more commercial subjects such as solid-state physics and the like. Even as the Physics Department retreated in its support of our program, the Mathematics Department remained more favorable. A few of the power-players in our department, notably a distinguished number theorist and a well-known topologist, made sure we would have no new hires in mathematical physics, and they would have liked to kill the program. But there was nothing to kill. We had no budget at all. We were doing it gratis. That is also the case most of the time for pure mathematicians. So in that sense, I am among comrades.

I felt extremely lucky to get the Boulder job in 1968. I knew the great roaring academic job bull market of the 1960s in the United States was rapidly shutting down. So I went through a few motions of looking for jobs in 1968, but I knew I would accept Boulder if an offer came. Virginia Tech in Blacksburg, Virginia, wanted to hire Seymour Goldberg from the University of Maryland and created a high-salary (about 50 percent more than I took at Boulder) position for him. But good old Seymour, who had taught me unbounded operator theory at Maryland, had no intention of leaving. So it was made known to me that I could have the Virginia Tech position. Also, Herbert Hauptman at NRL offered me a staff mathematician position there, with 50 percent of my time totally free for pursuing my mathematical interests. (As may be known to some readers, Herbert Hauptman went on to win the Nobel Prize in Chemistry.)

Minnesota had countered CU's offer by telling me I could expect tenure after one year. But my wife and I were psychologically already on our way back to Colorado. A friend from graduate school already there, Wes Wilson, had hosted a small party at his house while we were out skiing for a week in Vail. I did not even give a talk. I can still remember going into Stan Ulam's office at CU. It was the first time I met the man. He was sort of lounging in his chair at his desk, with that engaging and mischievous smile often creeping into his face. We chatted about mathematics and physics and after awhile he said, "You know, you can have the position if you want it." I told him, thanks, I would probably take it, but I needed an official letter stating the terms. I quickly received such letter back in Minnesota from the associate chair at the CU Mathematics Department. And after a very short negotiation, I accepted the job.

I learned later that I had been ranked #2, behind my fellow colleague Paul Fife at Minnesota, and ahead of Keith Miller at Berkeley. But Fife preferred Arizona and got an offer there. And Ulam wanted a mathematical physicist.

I thought I had been just "lucky," but I can see now that Stan knew what he was doing. The three hires here that year were Arlan Ramsey, a student of George Mackey at Harvard, Walter Wyss, a product of the famous mathematical physics tradition in Zurich, and me, who we may say was a (postdoctoral) student of Josef Jauch in Geneva. Stan hired three young mathematical physicists.

That I was also in partial differential equations was probably incidental to Stan. But he had not bothered to consult Watson Fulks, the senior PDE'er (that's how we abbreviate partial differential equations) in the department. I learned later that Watson was miffed, and for that reason he hired his former student Duane Sather the next year. But instead of undercutting me, Duane and I got along well and wrote some good papers together on nonlinear partial differential equations. And Watson may have undercut himself, for now there were two, rather than one, new PDE'ers.

Of course mathematical physics and partial differential equations overlap as domains. Much of physics is described by partial differential equations. In particular the Schrödinger equations describing atoms are partial differential equations. Partial differential equations relate positions, velocities, accelerations, and interactions. So they describe a lot of physics and engineering, in particular, quantum mechanics and fluid dynamics. They also model the futures and financial derivatives markets, an area of interest to me the last 15 years. Financial derivatives are indeed partial derivatives. So PDEs are a good tool to know.

Minnesota in 1966 was one of the very top schools in partial differential equations and that is why I had been advised to go there. Thus at Minnesota I was in the in-group, i.e., that particular subfield of mathematics that controlled the department. Many departments are that way. At Colorado historically it was Number Theory. I have come to call such in-groups, "the mafias." At CU I would never be in any of the controlling mafias. But when hired by Stan, I did not know that. Stan was chair from 1967 to 1970 and we managed to get the Mathematical Physics Ph.D. officially approved and off and running. With some help, I managed to keep it alive for thirty years.

But when Stan retired as department chair, the knives came out against applied mathematics and mathematical physics. The Administration had forcibly combined the old Pure Mathematics Department and the old Applied Mathematics Department in 1965 to form one large department. Ulam and Bob Richtmyer had been brought in from Los Alamos to provide some world-class noblesse-oblige personalities to the department. Six of us had been hired on federal monies, an NSF Center of Excellence grant. Almost immediately after Stan stepped down as chair in 1970 the resentments surfaced. In rather short order, ten younger members of the department who had interests in anything but pure mathematics were fired. That included Walter Wyss of our mathematical physics group; but I managed to get him transferred over to the Physics Department where he obtained tenure.

The fired ten included assistant professors with interests in Mathematical Education (of secondary school teachers) and in Computer Science. Computer Science as a field of science started out, almost without exception, within Mathematics Departments in the 1960s. By 1980 most of them had been chased out of the mathematics community. To survive and grow, they had to form their own departments and leave. I consider this one of the great errors of the American mathematical community.

Statisticians in general fared no better, although there are still a few departments around the country where they are welcome. Our two younger statisticians were fired in the house-cleaning. The two older ones, already tenured, stayed but were treated as untouchables.

Watching all this take place here in the early 1970s, my good friend John Maybee (who like me had broad competence in mathematics) and I formulated the following axiom: *The Abdication Principle: Pure mathematics departments by and large, will tend to retrench and narrow their vision to the most pure and abstract parts of mathematics.*

So it has gone here. From a department of about 54 members when I arrived in 1968, we are now about 27. Many synergies have disappeared. In face of this departmental refusal to respect the applied mathematicians in the department, in 1990 the administration split the department into a Pure Mathematics Department and an Applied Mathematics Program. Without going into detail, let me say that I, with help from others, had already succeeded in reinstalling the Applied Mathematics Ph.D. here, in 1980. But the pure mafias always gave it no resources. Some were happy to see it go.

An outside director for Applied Mathematics was brought in and that program eventually became a department in 1996. At its inception in 1990, six of us were invited to go over. Three did, including my friend John Maybee. Three of us did not. I am the only one left. The Dean put very heavy pressure on me in June 1990 to make the transfer. He threatened me with loss of salary, loss of Ph.D. students, and loss of courses. I fought back, telling him that I had Ph.D. students in all three of our Ph.D. programs (pure, applied, physics), and I spoke to him of illegalities in the formation of the Applied Mathematics Program. For example, I was asked to "transfer my tenure" over there. But that program was not an approved tenure-locus. There were other irregularities. Our Applied Mathematics Ph.D. was essentially forged away.

The Dean was right, though. I have lost a lot of salary, notwithstanding the fact that I have since emerged as the most prolific mathematician in terms of published papers in the history of Colorado mathematics, all universities, all departments, pure or applied. I could estimate: $1 million lost. I had no new Ph.D. students from the split in 1990 until 2000, after which I took two who were interested in mathematical finance. I lost all advanced graduate courses in partial differential equations and applied mathematics. But I have prospered nonetheless.

How could one prosper in such an environment? There are a number of factors that occur to me now. I say now because it seems in my nature to just continue on, shall we say, and follow my star, through good and bad. And I am a stoic, as was revealed to me early in some of our youthful mountaineering endurance tests. So I never really asked myself or inquired into the possibility of failure as betrayals fell upon me, not only at the university. One just continues. It helps if your missions are real and heartfelt.

But as I reflect, I can see that besides the advantage of my natural perseverance, I had a good running start and strong momentum. Certainly I was "protected" the first two years while Stan Ulam was department chair. Also, before the split in 1990, I had had since 1965 almost continuous summer research funding from NSF. Then there was the big NSF $22 million Optical Computing Systems Engineering

Research Center, which I helped CU win, and which funded me during the summers from 1988 to 1992. I was the only mathematician in that seven-department effort. After the 1990 mathematics split, I never received another individual NSF research grant. Who would fund a "half-breed" who refused to join the new applied mathematics unit? I quickly stopped applying. The wonderful Optical Computing Center grant was my last from NSF.

Then there was the freedom to do whatever research I liked within the pure mathematics department after the 1990 split. Pure mathematicians like to be left alone. In that sense, I am one of them. But I have an additional advantage. I can choose to do pure mathematics, applied mathematics, or mathematical physics, turning to each as the mood strikes.

What else? I found I could produce more research of the pure mathematics variety when I did not have Ph.D. students to train. In the engineering and applied mathematics research, Ph.D. students are an asset, and actually contribute to the team effort. In pure mathematics, one has to spend a lot of time training them.

No doubt another valuable ingredient in my career was my honorary membership, and coworkers and some travel support, in the Solvay physics institute in Brussels. Also we augmented that with some small NATO research grants. I helped Solvay get a big (1.5 million Euros) Neurocomputing grant. The Solvay Institute was the only mafia type connection that I have enjoyed during my career. But it endured for twenty-five years until the Nobel Laureat Ilya Prigogine died in 2003.

Stan Ulam left Boulder in 1974. I was really surprised to hear the attacks, or at least, the bad-mouthing of Stan here in the period from 1970 to 1974. The top powers in both number theory and topology disparaged Stan. The powers that be don't want another power to be. A younger new member of the department, from a family of great wealth and eastern U.S. private schools, criticized Stan's teaching and current research interests. It was true that Stan was considered over the hill compared to his earlier great mathematical and scientific achievements. Also due to his wide interests sometimes beyond mathematics, for example biology and neuroscience, Stan did not fit into any specific departmental or national mafia. The only grant he got here was one I and another younger mathematical physicist had obtained from NSF, to get visitor money for the new mathematical physics program. And not withstanding what he says in his autobiography, *Adventures of a Mathematician* (S.M. Ulam, New York: Scribner, 1976), I don't think Stan really liked teaching. In his book, he rationalized this in terms of the Polish education culture of no-need-to-prepare. In any case, he must have felt that his usefulness here was over. He was vulnerable and attacked, and so he left.

Bob Richtmyer, elegant and sophisticated and as Stan says in his autobiography, hard to really get to know, stayed. Bob really liked teaching and had strong feelings about it. Richtmyer helped me a great deal with the mathematical physics program; we more or less co-chaired it in the 1970s. I grew quite fond of Bob. My second wife Rose died in March 1980. I took her son and mine to Europe for two months summer 1980. When we returned, Bob's second wife Jane had already died of ovarian cancer, only six weeks after it was diagnosed. Bob was born on 10/10/10 so he was already almost 70 years old, whereas I was only 45 when Rose died. Jane, like Rose, was loved by all. Bob, like me, survived it and found life still enjoyable. After he retired, he still kept an office here and mentored many students, including some from local junior high schools.

As of this writing, I am still going strong. I thought about retiring at age 65, then again at age 70. And again more recently at age 75. But I have unique contributions to continue to make to the department and to world mathematics and science. In some sense, even though I have no power and no money, I am now an appreciated "éminence grise" (grey eminence) within the department. As one of the middle-aged group now running the department told me recently, "You have outlasted all your enemies, Karl." From the mathematical-physics side I am in a strange way the heir to both Stan's noblesse oblige and Bob's computer pioneering and books. I have 270 papers, Stan had 150. I have more books than Bob, although he was certainly the more polished writer. Because I teach the more applied courses, I am not in the competition to teach the more coveted pure courses. I actually feel welcome and appreciated in the department, even if, or perhaps because, I am not on most of their radar screens. But neither were Stan or Bob. I am in good company.

Probably I was the last member of the mathematics Department to see both Stan Ulam and Bob Richtmyer before they died.

On April 3, 1984, I gave a lecture at the Los Alamos Center for Nonlinear Studies, on explosion equations. My then-girlfriend Julie Inwood flew down to New Mexico with me so we could do some skiing afterwards in Taos. Then we returned to Santa Fe and I called Stan to see if he and Francoise happened to be in their home there. They were, and Stan said, "Come on over for some tea!" It was always so relaxing to be with Stan. He certainly enjoyed our visit and of course cast an admiring glance or two at Julie.

I was to give lectures on the explosion equations in Köln, Germany on May 16, and at the INRIA Laboratories in Paris on May 24. Just before the latter lecture I was to meet my French colleague Roger Temam at College de France in Paris, to hear a lecture by Peter Lax. Roger walked up to the restaurant where I was having coffee while waiting for him, and said, "Stan Ulam has died." Stan was 75 when he died on May 13, 1984.

With Bob Richtmyer it was quite a different story. He survived quite well in Boulder after his retirement. He had lady friends, enjoyed classical music, and for a while prepared meticulous lectures on fluid dynamics, which he gave for free at the university. Eventually, in his late 80s, he broke a hip. We heard he was being cared for by his "hippie" daughter somewhere in southern Colorado, near the town of Gardner. I presumed that I would never see him again.

But my climbing friends Bill Bueler and Norm Nesbit and I went down to the Sangre de Cristo mountains to climb Mt. Humboldt, a 14,000'er we had done long before. We summited Humboldt on August 21, 2002, and had dinner in Westcliff to plan our next day's drive down to the Spanish Peaks, where we would climb the day after. I said, "You know, an old friend of mine is down here somewhere; why don't we try to look him up?" All agreed and at the post office in Gardner the postmistress answered my inquiry with, "Why, I am sure Dr. Richtmyer would love to see you." Her directions took us out a little dirt road until we found the small, rather primitive-looking house of his daughter Roberta and her husband Owen. We walked in and Bob exclaimed, "Karl Gustafson!" I am pretty sure that the postmistress called Bob's daughter and told her my name, but you would not guess that from Bob's always sophisticated demeanor. He was hooked up to oxygen and in a wheelchair, but just beaming and conversant. His daughter told us that he could no longer remember any of the seven languages he once knew, and had even forgotten some of the books he had written. But there he was, beaming, the most sophisticated man I have ever known, spending his last years in a hippie house out in nowhere. All four of us, Bill, Norm, me, and Bob, are authors, which provided a common bond. Bill and Norm always fondly remember our visit with, as they referred to him, Mr. 10/10/10, the date of Bob's birth. Bob died about a year later, on 9/24/03, not quite making it to 10/10/03.

Notes

We in the department were surprised to discover in 1990 that our Ph.D. in Applied Mathematics no longer existed. That fact emerged when a secretary checking the university catalog for the next year saw that it now appeared only under the Applied Mathematics Unit's programs – which were not an officially approved part of the department. Such tampering with state documents is a forgery and a felony. A CU-Boulder committee formed to examine the case, and concluded that university administrators had violated academic rules. But no action was taken.

The reality that I have become the most prolific mathematician in the history of Colorado mathematics seems to have been a well-kept secret. But a recent external program review of the department brought it undeniably out in the open. Shortly after the report was issued, an unhappy colleague mumbled tritely to me, "But most of those were joint papers...." But I had already "done the math" and promptly replied: "Well, I am sole author on 160 of them, there was a co-author on 80 of them, and two co-authors on the remaining 30. So simple division and then addition would indicate: $160 + 40 + 10 = 210$ papers net." He looked stunned. Then I laughed and let him off the hook by saying, "I know you work damned hard on your research too."

7. The World Opens

... a community of scholars ...

Bribing My Way Through the Iron Curtain

It was May 17, 1972, and as we approached the first line of barbed wire separating the border of Austria from that of Hungary, the taxi driver I had hired in Vienna looked at me apprehensively and said, "I hope this works...." The Austrian border police looked at our passports and told him in German, "Good luck!" and waved us through into the no-man's land between the countries. As I looked left and right I saw foreboding Hungarian military watch towers and machine guns pointing at us as we drove about a mile to the Hungarian side of the border. After a short discussion with the Hungarian officers, my driver looked at me apologetically and told me either we both must go back, or he can drop me here and I can hope to get a visa at the border station. I hopped out and waved him goodbye as I proceeded to the border.

I was thus stranded at the small Hungarian border village of Hegyeshalom. I had had been invited by the great Hungarian mathematician Bela Sz. Falvi Nagy to give a lecture in his university in Szeged in southern Hungary. To get there, I was supposed to be on a particular train from Vienna to Budapest. The plan was that at my arrival time in Budapest, a mathematician named Janos Bognar would identify himself at the railway platform by waving a copy of the mathematics journal *Acta Szeged* over his head. But somehow I had neglected to get a visa to enter Hungary and go behind the Iron Curtain for the first time in my life. Visas would not be issued on the train. So I had been advised at my several-hours layover in Vienna to try to hire a taxi to pass through the border, and still get on that train when it stopped briefly in Hegyeshalom. Visas for those traveling by automobile into Hungary could in principle be obtained at border stations.

I joined the queue at the Visa Window but it moved at a snail's pace and I needed to catch my train at 1 p.m. and it was obvious that I was out of luck. Chatting with a German tourist waiting in front of me, without hesitation he told me, "You have to bribe the Tourist Officer," and pointed to the appropriate office off to the side of the building. I wandered over there and saw a pleasant looking middle-aged man behind his desk. He bid me enter and I smiled and explained: "I need to get on the train to Budapest at 1 p.m. when it passes through here. So I need a visa now."

"Sit down, please."

"Well, I'm sorry, but I don't have much time. I need a visa."

"Yes, you need a visa, but you also need a taxi to drive you to the railway station."

"Okay, but first I need the visa, and then the taxi."

"No, first you need the taxi, and then you need the visa."

"Yes, you are right, first I need the taxi, then I need the visa – and do you happen to know any taxi drivers who can come right now – and how much would such taxi cost me?"

"30 dollars."

"Done. I have 30 dollars right here," I said as I handed him the money. He disappeared and within minutes returned with the young lady from the Visa Window and immediately a taxi pulled up outside and we all got in and the tourist officer gave $10 each to the other two and I was dropped at the rail station with 30 minutes to spare. I stood there, all alone, in cold bright sunshine until some Hungarian soldiers came over, looked at me, my American passport, the just-issued visa, shook their heads, smiled at each other, then at me, and walked away.

I managed to see Bognar waving the *Acta Szeged* journal at the rail platform in Budapest and we walked for hours throughout the city. Almost no cars were present and everyone was walking, walking, walking. He deposited me at the Hungarian Academy of Sciences to spend the night. I was to give an improvised lecture around 11 a.m. the next morning and then be put on a train down to Szeged. Bognar could not be there but I found the small institute where the director welcomed me. We chatted in limited English in his office with his associate director. As the clock hour hand moved close to 11, I noticed a bottle of vodka and three glasses sitting in front of us. Suddenly it occurred to me that there would be no lecture unless I took the initiative and proposed a toast. Immediately the director jumped up and poured three large glasses and we three clicked our glasses together and in one motion drained them. Quickly we walked to the small lecture room. They were my audience of two persons and I can still see them semi-passed out but smiling as I mumbled and reeled in front of them. I had just given my first lecture behind the Iron Curtain.

It was a different story in Szeged. Sz. Nagy and his students were very interested in my recent results, which showed mathematically a limit of 118 for atomic numbers. Even now in 2009, experimenters are claiming they can (fleetingly at best) produce chemical elements possessing atomic numbers of 114 and 116 in their laboratories. No one has reached 118 yet, although they are getting close.

The Rathaus in the DDR

My next penetration of the Iron Curtain came in September 1977 at a large conference on Operator Algebras and Theoretical Physics organized in Leipzig in the DDR. That was the Soviet controlled East Germany. Many famous Russian mathematicians were supposed to attend. However, already the first day of the conference found many cancellations and substitutions to the program. It was explained to me that numerous Jewish Russian mathematicians were denied permission at the last minute. Their speaking slots were filled by lesser but more politically acceptable mathematicians. This was my first encounter with the Soviet political culture.

A brilliant East German mathematician named Eberhard Zeidler organized a special dinner at his flat. I and one other American, Colston Chandler, were invited. Also invited were two Russians, a powerful mathematician named Yurii Makarovich Berezanskii and someone else. Also there were two Brits and two Poles. The equation of eight was obviously an acceptable political balance. Somehow Zeidler had learned to manage the system and we had different elegant courses in different rooms throughout his large apartment. Berezanskii and the other Russian asked Colston and me many questions about our Western culture, educational system, all that. But I noticed that when we reflected the discussion back to them, they would not really answer us.

The party went on until about 3 a.m. and as Eberhard escorted us back to our hotel, he and I both had to pee. As we stood there peeing, I still remember him saying to me in his powerful deep voice, "You see, they do not even trust each other!"

On a free afternoon, Colston and I wandered down to the Rathaus (city government building) in the center of Leipzig. The first floor was typical of all Mayor's buildings but we had been told to take a look at the second floor. We were astonished to find the whole floor filled by high quality exhibits dedicated to the comradely connections between Karl-Marx-Universitat there in Leipzig and universities and institutes in Moscow and elsewhere in the Soviet Union. The two organizers of our conference, Professors Uhlmann and Lassner, were prominently featured, with photographs of them in Dubna, the Soviet version of Los Alamos.

Lassner was allowed out to participate in our VIIth International Conference on Mathematical Physics which I co-chaired in Boulder in 1983. I did not see Uhlmann again until a conference on quantum mechanics in Sweden in 2009. Clearly, both of these gentlemen had to make compromises during their careers under the Soviet domination of East Germany. Yet in 1977, 1983 and 2009, our mathematics was much more important to all of us than was the political system we each had to navigate. Mathematicians are a world-wide community of scholars.

Brahmins in Brussels

There we sat, Baidyaneth Misra, Kalyan Sinha, Jagdish Mehra, and me, in an elegant restaurant in Brussels. I had come to spend a month in the fall of 1978 to work with Kalyan, who was spending the semester at the new French-speaking university at Louvain-la-neuve a few miles south of Brussels. Baidyaneth and I had written several papers when he was on the Physics faculty in Boulder from 1970 to 1975 before he went to the University of Texas and then on to the Solvay Institute at the ULB university in Brussels. This was my first encounter with Professor Jagdish Mehra, who was visiting the recent Nobel Laureate (Chemistry, 1977) Ilya Prigogine at the Solvay Institute.

Mehra was a large fellow, and dressed to the nines. From brilliantly shined shoes to vest and fine-tailored suit, he strode about tapping his stylish cane and finally sat down again, expressing displeasure at the fact that no waiter was standing beside the table to be at his beck and call. Finally our first course arrived and lunch began. Notwithstanding his haughty manner, Mehra was on a mission, a mission to which he would devote his entire scientific life: to write the history of quantum mechanics. To do so, he had embarked on a program of collecting as much oral history as could still be gleaned from those pioneers of quantum mechanics who still lived, or from their close associates who still lived, or from their students. Baidyaneth, Kalyan, and I had all been Post-docs with the renowned Professor Josef Jauch at the University of Geneva. Jauch had died suddenly at age 59 four years earlier. So, with some interest tinged with disdain at our lowly status, Mehra plied us with questions about what Jauch had thought about the true meanings of quantum mechanics.

Of course his main target was Ilya Prigogine, who had known personally some of the great ones. The Solvay Institute had hosted the most famous series of conferences on theoretical physics of all time, starting in 1911. I have a wonderful collection of some of the original photos of those meetings given to me by Prigogine, and as I thumb through them now, I see him present already at the Huitième Conseil de Physique held at the Solvay Institute in 1948...also in the 10th Conseil in 1954. I would be privileged to participate in the 21st and 22nd Solvay Conferences in Japan in 1998 and in Greece in 2001.

Baidyaneth Misra and Kalyan Sinha were also Brahmins, but interestingly, with personalities very counter-posed. Baidyaneth was very proud, so much so as to cause himself professional difficulties. Not obtaining tenure at the University of Rochester, he was brought to the Physics Department in Boulder in 1970 as we built the Mathematical Physics Ph.D. program there. Baidyaneth was wonderful to work with one-on-one and he and I wrote four papers together. But he did not mesh into the culture of the Physics Department. In particular, he just refused to teach any courses other than the same advanced theoretical courses he taught year after year. So he did not obtain tenure in Boulder. However, he met Prigogine in Texas and Prigogine recognized his talent and took him to Brussels.

Kalyan on the other hand was a down-to-earth scrapper and fighter and very open and genial. I was able to bring him to Boulder for a semester in 1980 during which we completed a paper on quantum scattering theory together. With no hesitation I recommended him to a position at the Indian Statistical Institute in Delhi, India, where he had an illustrious lifetime career.

Mehra, as a historian of science, never found much support from the physics community, and according to Prigogine, was always asking for funds as he traveled the world on his mission. Before he died, he did succeed in publishing a unique and valuable many-volume contribution to the history of quantum mechanics.

Curfew in Warsaw

The spring of 1982 found me again behind the Iron Curtain, this time at the famous Stefan Banach Center in Warsaw where I gave five mathematics lectures. Due to recent political uprising by the Poles, the Soviet dominated Polish government had imposed an 11 p.m. curfew over the whole city. We were at a small party at one of the Polish professors' residences far out in the outskirts of Warsaw. At about 9 p.m. our host collected us in his living room and advised us that the party had to be declared over and everyone was to leave to be sure to be back at their hotels in the city center before 11 p.m. To reinforce this imperative to leave now, he told us the following anecdote.

It seems that some visitors to Warsaw were out too late and as the streetcars were moving too slowly, one of them looked at his watch and quickly calculated that he could not get to his hotel by 11 p.m. He told his colleague, "Look, you can stay on these infernally slow streetcars if you like...but I'm getting off and going to run for it." Shortly later as he was jogging across an intersection, one of the two young Polish soldiers there panicked and shot him. Nervously the two young soldiers approached the fallen tourist and found him to be dead. The one who shot him searched the body and found the name of the hotel. He looked at his comrade and said, "It's okay. He would not have made it there before 11 p.m. anyway."

I was again at the Banach Center in Warsaw in 1991, at another hard time as the Polish economy was in collapse. I, and others, would buy bananas to sustain ourselves, these being sold on every street corner, an innovation of some enterprising new Polish entrepreneur.

It is sometimes stated that Poland's history and fate could be largely explained simply by its unfortunate geographical position of being juxtaposed directly between Germany and Russia.

Dancing by the Black Sea

Sherman Riemenschneider, an American mathematician, and I had completed our presentations at an international conference in late May of 1984 in Varna, Bulgaria, at a lovely resort center created for Russian scientists on the Black Sea. The Bulgarian hosts were extremely friendly to us, as Americans did not often visit. Because many did not speak English, they even arranged that my lecture be given in French in a special session of presentations in French to accommodate some of the North Africans present. Sherman and I wandered down toward the sea that last evening and found a small pavilion and music and a table with about twenty of the Bulgarian students attending the conference. They beckoned us to join them, and Sherman, of football player physique but also a sharp mathematician, thought quicker than I and said, "Let's do: but first let's order some peach brandy bottles for the whole table!"

Immediately, two lovely young women invited us to dance with them. The quick-thinking pavilion owner then put on some American music. My partner was Denka Kutzarova and she quickly adapted to the music, excitedly exclaiming, "We do rock and roll!" We danced long into the night in loose and tight embrace in the pleasure of soft warm breezes off the Black Sea. For a number of years after, Denka sent me her lovely poems. Here is one:

> Winter Song
> The summer days all die in autumn
> and everything could be forgotten.
> The summer bird – it couldn't sing,
> but you remember still and think,
> you think of it – as if I've heard
> a winter song of such a bird.

Is poetry so different from mathematics? Both speak from the soul.

Paris...8 Times

It was a bitingly cold grey day in January, 1987, as I left Jill at the Hotel Royal Magda near the Arc du Triumphe and made my way down the Champs-Élysées to give my lecture at the First International Symposium on Domain Decomposition Methods for Partial Differential Equations. I had been invited at the last minute as a replacement speaker, all expenses paid, and as I began my presentation, the large audience of some of the world's best aeronautical and electrical engineers was at first curious about who this mathematician might be, could he really get any meaningful results? Then I could sense them sitting back in their seats, in mild resignation, just letting me continue...until I put up our slides of seven lift-and-drag vortices captured mathematically over the top surface of an airfoil, whereupon I can still hear the excited hum of the crowd, the top balcony straining forward in their seats to see the details of our results. I followed with our recent simulations of dragonfly flight, dragonflies obtaining six times the lift of conventional airliners, and the lecture was an unequivocal success.

A Paris Metro strike had crippled the city, although in French fashion a few lines were permitted to run. At the end of the day I made my way back to the hotel and Jill. Our chaotic brief marriage was falling apart and we were no longer living together, but at the last minute she wanted to come to Paris with me. I hastily called my colleague and friend Roger Temam in Paris and he quickly made a reservation for us at the Hotel Royal Magda. Jill was on her best behavior and we enjoyed very warm times in that wonderful hotel no matter the bitter cold outside and the Metro strike. I still remember a nearby restaurant called Le Relais D'Anjou where we had a long lunch one day. Across from us was a large well-dressed elderly gentleman and an attractive young woman who was perhaps his mistress. She was quite animated and carried at least ninety percent of the conversation, at times looking expectantly at him. He would respond by looking affectionately at her and say, "Je suis d'accord" or the equivalent, then she would continue. Jill was very impressed, would squeeze my arm, and tell me that was love.

There is indeed something about Paris that can bring out romance. I have been in Paris eight times, many of those visits involving invited lectures at conferences. My last visit was at the end of August 2006 after a conference in Belgium. I had dinner with my old friend and mathematician Anne Marie Boutet de Monvel. Long before, Anne Marie had loaned me her sumptuous apartment in Versaille, even her car, for a month in 1978 when I was on sabbatical. As I am reasonably fluent in street French, I always enjoy a visit to Paris. Moreover there seems to be something inherently intellectual about that city. One never tires of doing mathematics, or engaging charming company, or writing, in almost any of the myriad restaurants, all with fine food and pleasant atmosphere.

Dubrovnik and War

As I flew down toward Yugoslavia from Vienna in June 1990, sitting in the seat next to me was a lovely coed of a joint American-Yugoslavian family, returning to Dubrovnik from UCLA. As we chatted, she lowered her voice and in confidential tone told me that war would be coming, and why. I was a bit mystified, but the guards, who rather roughly shook us down as we departed the airplane at the Dubrovnik airport, added credence to her admonitions. This was several months before the ethnic war in which Yugoslavia disintegrated in fact broke out. Where was the U.S. Intelligence community?

At this International Workshop in Analysis and Its Applications, I wanted to get into print through its proceedings a rather fundamental result that my theory of operator turning angles showed for the first time the true geometric meaning of a famed inequality due to Kantorovich, a Russian who had been awarded the Nobel Prize in Economics. War broke out. I waited. Somehow, the editors did manage to publish the Proceedings, albeit with extremely limited distribution.

I cited that Proceedings paper and its mathematical result in another paper I submitted for publication in 1993. A couple of months later I received a very confidential fax from a rather secretive-sounding commercial agency, requesting a copy of my Dubrovnik paper, and offering to pay me a sum of $65. I was to send the copy of my paper to the commercial firm's address. They stated that they were to get the paper from me for some client who wished to remain unnamed. Should I wish to negotiate the price somewhat upward, that price could be considered. This was without doubt the strangest "reprint request" I have ever received. Of course, I did not comply.

Laundering Money in Colombia

In August 1990, before I left to fly to Cali, Colombia, to give a month of lectures, I told my son, "Don't send any money if I am kidnapped. Let the chips, or bodies, fall where they may!" It was a very bloody period in that country, essentially a state of total war between the drug cartels and the government, in which the leading presidential candidate Luis Carlos Galán had been assassinated the year before. Kidnappings were running at a rate of 100 occurrences per month. The warnings from the State Department were clear. Always be accompanied by someone who knows the locale. Always change your comings and goings, establishing no routine. From the moment I got on the Avianca flight from Miami to the time I got off the Avianca flight back in Miami, I did not see or talk to another American.

I had envisioned exploring the back-country of this beautiful South American country of two oceans, high peaks, and Amazon rain forest, but it was not to be. Too dangerous, my hosts insisted. Nonetheless, the first weekend my host and his family drove us north to Pereira where we spent the night and then the next day to the nearby hot and cold mountain springs resort of Santa Rosa de Cabal. After a couple of hours there I noticed two men closely watching me, even discretely following me, as I climbed up and down the cliffs between the springs and the swimming pool below. In my swimming trunks I am clearly a gringo, and no doubt they heard me speaking English with my host family. They went out to the parking lot where I saw them chatting, waiting, beside their four-wheel drive Jeep. It was late in the afternoon and we had to drive back to Cali so I alerted my host to the two men loitering by their Jeep. We quickly went by a circuitous route to our car and left. I have no idea whether I was in danger, but thereafter I followed the advice of my hosts and no longer sought to go alone to climb the high peaks.

Another weekend my host and I and the other principal lecturer, Patricio Aviles, a native Chilleno now in the United States, took the local bus south to Popayan. With rooster cages flying around as the bus labored up and down mountain passes and about 100 local stops and after passing several military roadblocks, after about five hours we arrived in that lovely white city. It was Sunday and ghostly quiet. After an elegant lunch we wandered up a steep incline to a hilltop where we could see families and children flying kites in the fresh breeze that gave relief from the heat of the city. From this hill we could see beautiful white churches in all directions. We decided to descend the easy way on the other side where there were steps. Patricio stopped to ask directions from a beautiful dark-haired lady in a pretty pink dress and white stockings, who was sitting on the edge of the hill. She jumped up lightly, called out sharply *"ninos!"* and with her two young children, walked down with us. At a sharp turn she lost her footing and I caught her elbow, steadying her. Her deep and different eyes flashed into mine for an instant and she asked Patricio about me. "Why are you here alone? Where is your wife?" she asked, Patricio aiding the conversation. *"Viudo,"* I replied, having learned this Spanish word for widower on the flight down, imagining that it might be needed. She missed a step, recovered, then shot me a look with even warmer eyes. *"Hay ninos?"* she asked. The conversation continued as we descended into the streets of the city. She was divorced, had gone back to school to study *la derecho*, the law. Nothing happens in Popayan, she went on, nothing at all; she had lived there all her life, in the most beautiful place in Colombia. She told us of the big earthquake in 1983, how Peruvians and others came to illegally claim some of the relief monies. A delightful day in Popayan, where nothing ever happens, except earthquakes and meeting beautiful, captivating women.

One weekend while I was in Cali my hosts decided we all needed a Salsa dancing night and we went to three different discos. The technique is to let the *aguardiente*, a wonderful local sugar-vodka, flow freely. Our first disco was Crystal, very plush. Soon we moved to a second, smaller disco called Coctail. The *aguardiente* was flowing, so were the women, and so was the Salsa. Your knees are always moving, but the arms do the pumping, your legs then following. It is said that Salsa gets in your blood and you get addicted. I would agree. There must be some kind of natural resonance between the rhythm and our brain rhythm.

Our last disco was Ricon de los Abuelos, in the heart of Juanchito, a black barrio. Salsa music poured from all the corner houses, and I was told that here the Salsa was the best in the world, better than the Cubans. One of my hosts, quite inebriated from the *aguardiente*, put his hand on my shoulder and said, "Juanchito. Most dangerous place in Cali. *Time Magazine* said so last month." I said, "Seems good to me!" and asked someone to Salsa.

The financial arrangements for the trip were that I would pay all my own expenses, including airfare and the apartment rental, and at the end I would be reimbursed, plus a small honorarium. I was aghast when my hosts handed me a paper bag, about the size of a small paper grocery sack, full of small denomination Colombian Pesos. I could just see me trying to explain in Miami that this was not drug money. One of my hosts said, no problem, he knew a local banker, let us see what we could do. We went to downtown Cali and a local bank. The kindly old banker chatted with his friend my host and after about half an hour, sent us a couple of blocks away to an indiscriminant looking address in a bland three-story building. A rather tough looking fellow answered the door and after some explanations were made, he led us down a dark hall and out the back and over to another building. We mounted a flight of stairs and he knocked on a door. It opened and I was astonished as we walked into a full scale bank, with tellers, desks, counters, everything you would see at the lobby of a major bank. The only difference was that there were no windows and entry was only through this simple second floor door. We were greeted by a young vice-president who promptly ran my bag of pesos through a counting machine and then issued me a cashier's check on a major international bank.

I was not home free yet, however. The morning of my departure, my hosts dropped me at the airport. At security I frantically searched my pockets and realized that I had misplaced the key to the padlock on my old Italian leather suitcase in my hurried early morning packing. Probably it still lay on the floor of the apartment where I had battled to get all my purchases stuffed into my luggage.

"*Perdito me llave*," I mumbled.

The Colombian chief checker spoke English. With an almost unbelieving but non-hostile smile playing on his face, "Lost your key?" he stated softly but in incredulous tone. I told him to just cut the lock to look inside. A line was forming behind me. We smiled at each other. He waved me on, with a "*Siga*".

"*Muchas gracias*. I'll have the same problem in Miami," I muttered, smiling.

"In Miami, yes" he agreed, smiling genuinely. But in Miami, after I identified myself as a professor of mathematics, the customs agent just welcomed me back and waved me through.

Russia in Despair

I looked all around the arrivals hall at the airport in Moscow and I did not see George Kobelkov, who at a conference in Warsaw the year before, had invited me to be his guest at Moscow State University for a month. It was May 1992 and Russia had slipped into an economic black-hole. I had been standing there already thirty minutes, it was 5 p.m. in the afternoon, and exactly what I did not want to happen – to be stranded upon my arrival, with no Russian language skills – was happening. I had no address, did not know where I was to stay, nothing.

Then I saw George. With him was a well-dressed darker-skinned man. I shouted, "George!" The frown and worry left his face as he called back, "Karl!" as they made their way over to me. He explained that he had never received my letter with my flight number so they had been meeting all flights. His companion was introduced as Husain, and he worked for the Syrian Consulate and had a car.

We drove into Moscow, past Pushkin Square, turned by the Kremlin, and headed southwest. Coming out into the Lenin Hills, I saw for the first time the huge Moscow State University building looming up into the evening sky. The tiny room I was placed in there had a small dirty toilet and no toilet paper. George explained that this room cost only 50 rubles a night and that the university no longer had any money, not even for his salary, and that it would cost a fortune to put me up at a downtown hotel. He and Husain looked at me very apologetically and I realized I would just have to make the best of it. "Don't worry," I said, "I see some old newspapers by the toilet." George brought me a small roll of toilet paper the next day, stolen from his own home's short supply. He also brought me a small electrical coil to boil my water taken from the faucet.

The reform economics had crashed the Soviet system and had produced 10,000% inflation, thereby reducing professors' salaries to $30 per month, for which they often had to wait months to receive. The inflation meant that the cost of my room was only 50 cents, including its cockroaches and filth. I like to have the real experience of a country, not that of plush tourist hotels, and I realized that indeed, I was going to experience such a month here at the new Russian poverty level.

I had arrived in Moscow just as the long May Day four-day weekend was to begin. George lived all the way across Moscow and would not be back for three days. The next day I found the Stalovayah, a cafeteria, but it was closed. For me, my first May Day in Russia would be a hungry one. However, as I continued to explore this massive university building constructed after World War II by German slave labor, I found the second Stalovayah and it was open. A long queue of students went down the stairs to the single booth with a glass window behind which sat an old lady. As I got closer, it dawned on me that the students had to squint through the dirty glass to try to make out the menu for the day and then ask for paper coupons to present to the food serving ladies. I knew no Russian. *"Po English?"* I asked the lady in the booth. She looked at me with contempt and a stream of Russian words came from her. I turned to the fellow behind me, whose features suggested he was Mongolian, pointed to my mouth, then to the menu hidden behind the glass, and gave him a few rubles. Looking at me in awe, his mouth fell open, I pointed again at the menu, shrugged with my palms up, and he got it. He ordered some coupons for me and gave me my change. As we walked into the cafeteria I said *spasibo* (thanks), and he motioned me into the food lines where I would trade my coupons. Getting our food slopped onto plates from huge vats, we made our way to the tables.

The food was not bad for someone who has lived as I have. After all, I had found Army food and hospital food to be just fine – it existed! So I wolfed it down, whatever it was. But I realized I would have the same difficulty for each meal for the next three days as this Stalovayah would be my only food source. So before my Mongolian-Russian new friend could leave, I went over to him, pulled out pencil and paper, convinced him in sign language that I would like to order exactly what he would order, and how do you say that in Russian? He wrote something in Cyrillic which of course I could not read. I motioned for him to say it and it sounded like "*Tosha samoyeh*" in my phonetics. I shall never forget that way to say "the same as he has" which I would use at the window for the first week until I could learn some Russian. I was happy to eat exactly what the person in front of me in line at the coupon booth would order!

I learned that almost all such institutionalized food during this Russian famine was coming from NATO and European Community surplus food bins. I very much liked the breakfast porridge called *kasha*, one variety of which tasted like our cream-of-wheat cereal. At dinner one could usually get *kartochka* (potatoes), and at every meal one could get a *salat iz ogurtsov* (cucumber salad). I survived the month but steadily lost weight.

The Stalovayah, although it sustained me, had a tragic history. In the early days of communism, the Stalovayah was one of the great success stories of that brave new society. Each workplace had a Stalovayah to provide a nutritious lunch to all workers in the country. The idea was that, conversely, then all would report dutifully to work. But in the years after Glasnost (the opening) and then Perestroika (the reforming), the Stalovayahs went into precipitous decline. I paid only 10–20 cents per meal there, but 50 cents a day for the average Russian in 1992 was too much. So I, and the students, were better fed than the man in the street.

During my visit, I hired George's 18-year-old daughter Tanya to give me many Russian lessons, and I paid her liberally, knowing my rubles would flow to George's family. A few times I took the Moscow Metro all the way across Moscow to George's apartment in Xumke (pronounced Himkee), a formerly secret military city like many that surrounded Moscow. George's wife Gala was terrific and I happily played chess with his ten-year-old son Sergei, often losing to him. This was a very nice family. George arranged a three-day rail trip up to St. Petersburg where we visited the fabulous Hermitage museum. We witnessed the May 9th 22-gun salute to the defeat of Germany, a tradition now being established to replace the old May Day ritual.

I gave two lectures, the first at a Workshop on Nonlinear Optics arranged by Yuri Karamzin, a dead-ringer for the actor Burt Reynolds. Afterward Yuri invited me to dinner at his apartment, where his gracious wife had prepared an elegant although sparse multi-course dinner. Both were highly cultured Russians several generations removed from the village mentality. Because of this I sensed in Yuri an inner turmoil that his less sophisticated colleagues would not have to endure. Yuri excused himself every half-hour for a cigarette outside.

My second lecture was about our new lift-dynamics results for hovering insects and hovering aerodynamics vehicles. This attracted a large audience at the university. I was very surprised to see many military uniforms in the audience, most of quite high rank. Afterwards a young general with full medal array approached me and in halting English described his own research to me. As we looked at some diagrams in his papers, he grinned at me and said "Stealth!" I smiled and told him one of my Ph.D. students now at Boeing had done a lot of Stealth design. The young general and I then just broke out laughing at the ludicrous nature of our situation. We laughed for several minutes, shook hands, and parted on very good terms.

One weekend day I had taken the Metro over to Xumke and after lunch George and I strolled in the lovely poplar woods nearby. George told me of the problems with "The Southerners," who were seceding from the Soviet Union. It seemed while I was in Russia that most Russians were in denial that such disintegration of the Soviet Union was really happening. Also George "missed Stalin." Most Russians I talked to thought of Breznev as a fool, Gorbachev as the worst thing that had ever happened to the Soviet Union, and Jeltsin as totally corrupt. George was of the opinion that Russians needed "a strong clever leader." I asked him, "Wouldn't a strong honest leader be better?" He thought a moment and then replied: "No. He must be, above all, clever. Russians have no experience with Democracy. We have no judicial system. We do not have a tradition of free enterprise. Our leader must be like a Czar, but more clever." This would explain the popularity of their current leader Vladimir Putin.

By chance I was reading the English language newspaper The Moscow Times on Friday, May 8. A headline on Page 7 caught my attention: American was Assassinated, Family Says. I read on. Martha Phillips, 43, a Trotskyist activist who had lived in Moscow since the preceding September, was found on February 9 stabbed and strangled in her Moscow apartment. Valuables were not taken. There was no sexual assault. Her sister, an attorney in Denver, was quoted as saying it was a political assassination. I did a double-take. Could this be Herb Greenberg's daughter Martha? Herb was a mathematics professor I knew at Denver University. Later I confirmed that indeed it was.

Moscow police claimed initially that Phillips died of natural causes, no matter the forensic evidence of stabbing and multiple bruises on her neck. The chief criminal investigator of the Krasnaya Presnya district, where the murder had taken place, refused to comment on any of the specifics of the case, except to say the apartment had not been broken into. Martha's boyfriend was not a suspect. The International Communist League in which Martha was a vocal member organized some world-wide protests. But nothing followed.

George and Husain one evening took me to two of the new nightclubs in Moscow, where pretty, naked young ladies pretended to be exploited, and to exploit each other, sexually. It was difficult for George and Husain to accept that I never go to such clubs in the United States. George had no car and depended upon Husain for all such transport. I developed the distinct impression that George also had close contacts with the KGB. He told me once that many Russian scientists needed to cooperate with the KGB, or at least, to remain on good terms with that organization and to have some friends there. When the time came for me to leave Moscow, some friend of George had a car to take me to the airport.

* * *

I was to visit Moscow again in six short months. Upon my return to the United States after my first visit, I was deluged with requests by various U.S. government agencies to tell them what I knew from my trip, and would I recommend any particular Russian scientists for the new little grants the NAS and NSF were trying to set up in Russia. The Soros Foundation plied me with Russian applications for their grants, which I knew most Russians viewed with suspicion. Apparently my visit had been just ahead of the curve as the depth of the Soviet economic collapse was only now beginning to be fully perceived in the West. The National Academy of Sciences more or less insisted that I accept their travel grant to go back to Moscow. Thus on January 5, 1993, I found myself again deplaning at Sheremet Airport in Moscow, met there by George, Husain, and George's son Sergei. As we drove into Moscow, I commented on how much cleaner things looked. I did not see what I had called in my earlier trip "The Standing People," lines and lines of Russians we had seen in both Moscow and St. Petersburg, standing on the streets offering their personal belongings for sale. As my NAS grant was paying my expenses, George had obtained a much nicer room for me at the university. Small kiosks had sprung up near the Stalovayahs and one could buy bread, cakes, and sausages there. My Russian colleagues grumbled, "Prices too high," but things were getting better. We celebrated the Russian Christmas, which they call just "Christmas Tree" on January 7. I gave the annual Kolmogorov lecture, named for the great Russian mathematician who put probability theory on a rigorous mathematical basis.

Mosque Demolition and Riots in India

Between my two visits to Russia, I was part of a lecture contingent brought to India in December 1992 for three weeks and six planned lectures. Part of my expenses were paid by a Mr. R. K. Roy, a book publisher in India whom I had met in Boulder and to whom I gave permission to publish a third Edition of my PDE book in India. My flight arrived in Calcutta on December 5, and at 6 a.m. on December 6 we left Calcutta to fly down to Bangalore, already becoming a major science city for India. At noon on December 6, a Hindu mob estimated at 150,000 strong broke through police cordons in the city of Ayodhya and, using only hammers to knock down the domes and then their bare hands to tear down bricks, totally destroyed the historic Muslim Babri Mosque. Hindu-Muslim riots exploded throughout India. Already that evening our bus to a banquet had to detour around a Muslin section of Bangalore, from which we could hear the sound of gunfire. The Indian scientist sitting next to me calmly remarked, "Well, they are shooting Muslims in there." More than 2,000 died across India.

K. S. Yajnik, the Head of the Computational division at the National Aeronautical Laboratory there in Bangalore, told me later that he almost cancelled the conference right away. But a lot of money had already been committed. We were banqueted in the evenings by Holiday Inn, Convex Computer, DEC Computer. It was surreal, the worst riots in decades were happening around the country.

The Ayodhya Babri Mosque had been built in 1527 following the destruction of a Hindu temple there which had ties to the birthplace of the god Rama. It is alleged that the BJP political party condoned and perhaps even planned the incident of December 6 in order to polarize the Indian public. Indeed at elections in the following years the BJP party rose to prominence.

The consequent collapse of our lecture tour became evident when we flew back to Calcutta at the end of the week for the S. N. Bose Platinum Jubilee Conference the next week. Almost no one attended and fear was in the streets. Our scheduled lectures in Delhi and the great engineering complex in Roorkee were cancelled. We became refugees. My old friend Kalyan Sinha called and told me to try to catch a flight to Delhi where he could put me up at the Indian Statistical Institute. The noted linear algebraist Beresford Parlett and I made our way to the Air India ticket office in downtown Calcutta, walking over those linen-wrapped dead bodies that families with no money just left in the streets. Beresford had a former Ph.D. student at the Tata Institute in Mumbai. "I'm prepared to bribe to get on a flight," he told me. The crisis was compounded by a pilots strike at Air India and managers were flying the airplanes. But the ticket officer gave preference to foreigners and both Beresford and I got our flights.

Kalyan found a driver and sent Vaughn Jones, a strapping New Zealander who had won the Fields Medal a few years earlier, a Frenchman Frederic Pham, and me, on a three-day tour, after which I was to give a lecture at the ISI. We set out on the highway south to Jaipur. It was a nerve wracking experience and Jones, Pham, and I agreed to rotate into the death seat beside the driver each hour. We made it to Jaipur where both the driver and Pham said they would go no further. The next morning Vaughn and I set out on foot and spent a wonderful day in Jaipur with its ancient astronomical observatories and pleasant location at the edge of the western desert country. An elegant Indian school teacher saw us on the streets and sent out his son to invite us into his elite girls' school. Vaughn and I were introduced to those startled pretty teenagers as scholars from a faraway land. Karma was good in Jaipur.

That evening Vaughn and I found a night bus to Agra. A few miles outside Agra the bus was forced to stop at a military blockade. Vaughn jumped up, looked at me, said "This is it!" and we prepared to fight. However a fleet of Indians with their little one or two passenger mopeds were at the ready and we bargained for one who took us into a hotel in Agra. The next day we had Agra, and the Taj Mahal, essentially to ourselves. We found a night train back to Delhi.

India had yet another surprise for me. Somehow my old friend and colleague Baidyaneth Misra, who had left Belgium to semi-retire back in India, had learned that I was in Delhi. Misra and his wife and their three lovely daughters were ensconced for six months in the guest hotel at the center of the Maharishi's Ashram about an hour outside Delhi. Would I like to visit there the last day before my flight home? A car was sent and I was astonished at what would greet me there. Arranged in concentric circles, the Ashram housed about 7,000 students. The Vice Rector, Dr. Gyanendra Mahanpatra, spent the day with us. We were taken to a room of 500 youths in yellow robes who would, at the clap of a hand, levitate. In their minds, they were hovering in the air, but to us watching unobtrusively from a corner near the door, they were actually flopping up and down. The rector told us most of these youths had been collected off the poverty-ridden streets of India and were being trained in meditation – so they were better off in the Ashram. The Maharishi had always sought help from theoretical physicists at his universities in India, Switzerland, and Iowa, and Baidyaneth was a guest under those auspices. There are possible although highly speculative connections between quantum mechanics and human consciousness. To my query to Baidyaneth, he told me that when asked his opinion about such matters, he just answered as best as he could, and honestly.

Mr. R. K. Roy, who was to publish the Indian version of my PDE book, took good care of us in Calcutta. I really liked this gentleman, a blocky, rugged looking fellow originally from Assam. He would send cars for us to get us to and from the conference, usually riding in one of them as his drivers negotiated the Calcutta streets bordered by continuous little shacks where whole families lived, sometimes their little babies just left abandoned in the street. Mr. Roy had only one child himself, a beautiful enchanting little girl named Bashiki. She was eight years old and often rode along with us. I sometimes wonder what happened to little Bashiki. Mr. Roy and his family just vanished the following year. I tried hard to get some information from his staff there in Calcutta, but they claimed that nobody knew. Speculations were that he was murdered over gambling debts or that he took the money raised for his book-publishing venture and went to hide in London. Eventually I obtained three copies of the Indian edition of my book, which never went into distribution.

China and into Tibet

As I disembarked the South China Airlines flight from Hefei to Tian on May 26, 1994, I anxiously looked around the terminal and was delighted to see two other tall Westerners, my long time climbing friends Bill Bueler and Norm Nesbit, also entering the terminal from their flight from Hong Kong. The timing could not have been better. I had earlier flown alone to Shanghai, caught a night train to Hefei where I gave two lectures to engineers and mathematicians over two days. After several hours delay at the train terminal in Tian, we all three managed to get on a night train heading west to Xining. Our goal was to drive overland from Xining to enter Tibet through its northern mountain ranges and high plateau.

But there were no seats for us. The conductor led us to a decrepit coach car where there was a single wooden bench on which two of us could sit. The rest of the car was filled with Tibetans, some sitting in circles eating, some trying to sleep on the floor. We stared at them, the first native Tibetans Norm and I had seen, and they stared wide-eyed back at us, the first Americans they had seen. These were very poor Tibetans, their faces and hands covered with grime. Bill, fluent in Chinese Mandarin, went to find some conductor who might find us at least some hard sleeper berths somewhere on the train. He succeeded and so eventually we smiled warmly at our Tibetan friends who smiled even more warmly at us as we waved our goodbyes. Not a word had been spoken.

We enjoyed a day at the Taersi Monastery (Kumbum) near Xining. Kumbum was the birthplace of Tsongkapa, the founder of the Yellow Hat Sect, the dominant one in Tibet. The present Dalai Lama was born near it. Historically this region was Tibetan – in fact about 25 percent of China is actually occupied by Tibetans, who however are now losing ground to the imported Han Chinese. We were in Qinghai province, northeast of and almost as large as what is now called Tibet. The population there is about 40 percent Han, 30 percent Tibetan, and 30 percent other minorities (Mongolian and Hui/Moslems).

With a Chinese driver and an official Chinese escort, it took us six days to drive up onto and through the Tibetan plateau and over two high mountain passes to Lhasa. During the first three days we saw hundreds of the tractor-pulled carts overloaded with three or four men and gear. These were gold miners heading up into the Kunlun Mountains. Most were Hui (Moslems). We found a large Tibetan festival near Quinhai Lake at 10,460 feet. This is the Amdo region of original Tibet and has a different dialect: instead of *"tasha-dilai"* for "hello" as in Lhasa, here it was *"toedemo"*. After a night there in a run-down guesthouse, we went over a pass to a semi-desert valley populated mostly by Mongolians who raised camels in this extension of the Tsaidam desert basin. Their yurts were surrounded by mud-walled enclosures for sheep and guarded by extremely fierce dogs. A night in the little oasis town of Dulan was followed by a long hot drive across the barren desert to Golmud. In this desolate Sahara-like landscape, the Chinese government has placed some of their "labor reform camps." Golmud is an ugly industrial town and was at the end of the railroad when we were there. Now it is the jumping-off point for the spectacular new rail line the Chinese have built all the way to Lhasa.

In Golmud, as we were set to continue the next day up onto the Tibetan plateau proper, we were met with a nasty political surprise. Our Chinese escort was told that neither he nor our driver would be allowed to enter the Tibetan Autonomous Region. Only because we three had already obtained permission would we be allowed to continue to Lhasa. The Chinese government was closing Tibet during the days surrounding the fifth anniversary of the Tiananmen Square massacre (June 4, 1989). Our escort Xiao Ren was told that 30 train cars of troops had been sent to Golmud and were then being trucked to Lhasa to deal with any trouble that might occur. We were to drive the next day to the tiny village of Tuotuohe just north of the Tibetan border, and wait there until a car with new driver and escort would be sent up from the other side to take us on to Lhasa.

The Tibetan plateau sits at about 15,000 feet above sea level and resembles South Park in Colorado – although it's 5,000 feet higher, much larger, and with no forested mountains at its edge. We saw hundreds of wild gazelles which the Chinese call *huangyang* or Mongolian gazelles, several dozen wild asses (kiang), hawks, eagles, foxes, and of course wild yaks and yak caravans. I was in a dream come true, for I had as a youth always yearned to see Tibet. Bill and Norm felt the same way. That night we arrived at Tuotuohe, the ultimate source of the Yangtze River. The guesthouses had steadily declined in quality and we spent the night in bitter cold at 15,000 feet in our little emergency sleeping bags we had packed for such situations. Norm continued taking photos of "the world's worst outhouses." Bill started having trouble with the altitude and could not sleep.

Our new crew arrived from Lhasa and the next morning we left in a snowstorm to cross Tangula Pass at 17,163 feet. I loved this barren high pass and have a photo of a small prayer-flag pile at its top in my living room in Boulder. I wanted to stay there and wander around a few hours but the driver and Bill's altitude difficulties cut short my reverie and we descended the less desolate windward side of the pass to spend the night in Nagu, still at 15,000 feet. The next day we finally arrived in Lhasa. The Holiday Inn there seemed like paradise! We would be in Lhasa from June 3–7, and almost the only Westerners there.

After enjoying the Potala, Jokhang Temple, Barkhor, and Sera Monastery, we had two free days. Our new escort, who had been obtained at the last minute, was not very experienced and since Bill was fluent in Chinese, turned us loose. Bill had always wanted to climb Mt. Gyembuwudze, also called Gompe Utse by Heinrich Harrer in his book *Seven Years in Tibet*. Our escort just called it Simi mountain (Holy mountain), the most settled-upon name being Mt. Gephel Ri. It is about 17,200 feet high, rises about 4,500 feet above the Drepung Monastery, and is considered sacred by the Dalai Lama. Bill had attempted it in his first visit to Lhasa but had to turn back at 15,000 feet due to altitude.

This time he made it, although two hours behind me to the top, one hour behind Norm. For some reason I felt great, probably because I had been sleeping well the entire trip. Even though this is a "trail climb," it was one of the high points in all our lives. We had the mountain entirely to ourselves.

In our last free day, I wanted to visit Tibet University, at that time a small campus in the center of Lhasa. It was open and in an empty classroom we found a blackboard full of the mathematics of complex variables – it would look the same anywhere in the world! We also found a Tibetan mathematics professor, Mr. Bai Ma Duo Jie. All literate Tibetans now of course speak Chinese, and the Tibetan professor had spent several years of his education in China.

We all three then flew to Chengdu. Bill negotiated hard with the Sichuan Tour company and secured a driver and escort for a three day drive up into the Wolong panda nursery above Chengdu, where we also took several hikes hoping, but failing, to see a wild panda. Bill and Norm then flew out to Hong Kong and home. I stayed in Chengdu to give two lectures, one to the Academia Sinica (Chinese Academy of Sciences) branch in Chengdu, another to an engineering school in Chengdu. Then I flew to Beijing where I spent some days and gave two more lectures.

I was so impressed with my old climbing friend Bill Bueler's perfect command of the Chinese language. Of course I knew that Bill had been a top CIA China-Tibet specialist in earlier times. But Bill, in return, was impressed that I could arrange to join them in a mathematical lecture tour surrounding our main objective, Tibet. Until Bill pointed out to me that my mathematics was now opening up the world to me (witness Russia and India the previous years) that transition to world-class mathematician had not formulated itself to me.

* * *

I have been told for many years that I would not even recognize the new China. But it was already emerging when we were there in 1994. And the breakout from Mao's iron grip had already begun in 1979, fifteen years earlier. Surprise of surprises, early in 2010 I would be invited to lecture in China again that year, and I would visit again while writing this book. Indeed, as I was told, the changes taking place in its cities are breathtaking.

* * *

I was to return to Tibet again in 2001, quite by accident. My colleague Kim Malville in the Physics Department and his wife had arranged to take nine hardy (e.g., triathletes, Nordic skiers) students to Nepal and then trek over the border to Mt. Kailash in Tibet. However, Kim's wife Nancy needed a hip replacement. By chance she saw me working out at a CU gym and asked me if I might like to replace Kim, their having already found a female replacement for her, a strong 60-year-old Physics professor named Irene Little. I had no obligations in the envisioned May-June period, so I said, "Why yes, of course!" I would be 66 years old.

The day before we left Denver to fly to Katmandu, the Chinese government cancelled our permit to trek from Similot in Northwest Nepal over the border to Mt. Kailash. Nonetheless, all eleven of us decided to go to Katmandu and see what our Sherpa guides might find as an alternate trek. What great luck was to befall us! After some days delay, we loaded into a bus and with a change to Land Cruisers crossed the border into Tibet to spend two days acclimatizing at the first Tibetan town of Nyalam, at 12,000 feet. The plan was to trek up a very remote valley (Kharta Chu) just north of Mt. Everest (Chomolungma, 29,200 feet), then traverse across trailless high country to meet a small pass (Doya La), descend this trail to the Rongbuk Valley, and walk up to the Rongbuk monastery and the Everest base camp (17,500 feet) on the north side of the mountain. This we did. It was extremely remote country, some of the terrain difficult even for the yaks to handle. For ten nights, we slept in our tents, between 14,000 and 17,000 feet. Our three days in the Rongbuk area offered me my first views of Everest, at first clear and spectacular in the day and in the moonlight, then shimmering in the sunlight after a night snow that covered our camp with almost two feet of snow. Rongbuk was practically deserted as the spring climbing season was over and the mountain closed.

Thereafter we became tourists and spent three nights in the wonderful Tibetan towns of Shigatze and Gyantze. After we had emerged from the remote valleys to Rongbuk, our Sherpas were deeply saddened to learn of the massacre of the Nepalese royal family (11 of them dead) that had occurred in Katmandu during our absence. When we arrived in Shigatze, our students went online to assure their parents back in the United States that we were all okay. After all, we had been in Tibet, not Nepal!

Much could I tell of this trek and this wonderful group of young students – it was a fabulous adventure. One vignette comes sometimes to my mind. After a hard climb up out of the Kharta valley, we put in a camp at about 16,000 feet and settled in for the night in soft-falling snow. The monsoon had started and was already creeping over onto the leeside of the great Himalayan peaks. In a cold wind blowing off the range, the next morning we made our way up onto a rock and ice glacier next to some not quite frozen desolate lakes just to our right on the tilting plateau at 17,000 feet. We were temporarily lost and huddled to consult our limited maps. Suddenly I was struck with the beauty of it all, the roaring biting wind, the vast expanse of glacier and lakes, the peaks above shrouded in grey clouds. I turned to Adam Kay, the student poet and writer among us, and shouted above the wind: "This is it! This is God's country! Can you believe how beautiful this is?!" Adam looked at me, then over at the lakes, then back at me as if I were crazy. But then he got it, "Right! Right, Karl! Right!"

The Country Club at Les Treilles

The driver was waiting for me at the Nice airport in late June, 1994. With few words, he escorted me to his private car and we drove west into Provence. A few kilometers beyond Draguigan, he turned onto a small country lane which turned and twisted up to the Grande Maison, a combined meeting hall and haute cuisine restaurant in which we would enjoy all our lunches and dinners for a week. An elegant Parisian, Catherine Bachy, welcomed me at the reception and handed me the keys to my car and directions to a private villa that I would share with one other participant. I had arrived at the prestigious Foundation les Treilles estate, established by the late Mme. Anne Schlumberger in one of the most beautiful locations of Southern France. In her words: "I imagined a mathematician whose mathematics were poetry, a poet whose lines would invent space, a musician bringing forth symphonies – a place where light, comparable to that of Greece, would prompt painters to paint, sculptors to sculpt." About six meetings of fewer than 25 participants would be organized each year. I would have the great fortune to be invited to three, in 1994, 1996, and 1999, concerned with mathematical physics.

Each morning we would awake to the sun of Provence and a fresh breakfast basket placed just inside the door of our private country villa. In the overwhelming fragrance of lavender we would leisurely drive about a mile to the Grande Maison where the informal lectures would begin. After a fine lunch, we would coffee for an hour on the terrace and then gather for the short afternoon session. There would be plenty of time for a walk through the oak forests and olive groves before dinner, which was always of the highest cuisine, served by waiters of impeccable standing. The atmosphere of these incredible meetings at Les Treilles was like that of a country club or posh exclusive resort, but was even better because, as I would put it to my friends, we invited participants were scientists and artists and not monied.

Nor, in general, were we famous, although top scientists were indeed plentiful among us. Professor Ilya Prigogine was at all three conferences. The great Russian mathematician Israel Gelfand joined us at the 1994 meeting. The Indian physicist George Sudarshan, often thought to be deserving of a Nobel Prize, joined us at the 1996 meeting. The controversial but brilliant Italian quantum probabilist

Luigi Accardi added a lot of life to the 1999 meeting. But the goal was always discussions, relaxed discussions on the terrace or in the lounge or at dinner or on petite promenades. There were bright young inviteés mixed with more established thinkers.

I really liked the candid manner of George Sudarshan and we had many interesting conversations at this and later conferences. Always a rival to Prigogine, George opened his presentation on the third day with, "I have noticed that in their thirty minutes here, none of the speakers has had time to get to their conclusions! Therefore I am going to state my conclusions at the beginning!" This he did, beaming at the small assembly, and waited...sure enough, Prigogine immediately attacked his conclusions, asking how he could possibly even think they could be true...point and counterpoint followed with a few others joining in...and George never got back to the beginning of his lecture.

Book Tour in Japan

Asia again called when the publishers of the translation of my PDE book into Japanese organized a small two-week lecture tour for me in late October, 1995. I would be accompanied by Professor Takehisa Abe, who had headed the team of three translators. We had met very briefly the preceding year at an IMACS conference in Lyngby, Denmark. Takehisa's command of written English is exceptional, with large vocabulary and almost flawless usage. But I had been astonished to find in Denmark that we often could not communicate at all in spoken English. Fortunately, when he met my flight at the Narita arrival lounge in Tokyo, he brought along in addition to his lovely wife Keiko-san, his vivacious daughter Akiko-san, who speaks French. Because I speak French, Akiko was a pleasant surprise and great help during my lectures in Tokyo. By the time Takehisa and I went down to Kyoto for my lectures there the next week, he and I had learned to navigate quite well via some English, a little Japanese, and a little note pad we passed back and forth.

My first lecture was to the powerful theoretical group of physicists at the University of Tokyo. Afterward we went out for sake and dinner, as is the custom in Japan. As the evening went on, two of the brilliant young Japanese physicists sitting beside me laughed, and one said, "We have a question for you, Professor Gustafson." I expected a really tough technical discussion about some mathematical physics issue but with no other recourse I of course gamely answered, "Fine! Go ahead!"

"Well, we wonder: Do you believe in God?"

I leaned slightly backward, relieved but confused, took a draught of some more sake, and then replied, "Maybe." "No," they said. "As a mathematician, you are supposed to say, 'Yes. Of course. But only up to unitary equivalence.'"

This was a very nice "in joke." Many results mathematical-physicists are able to prove hold in any coordinate system, which one may transfer among by what are called unitary transformations. But because the theorem holds for any coordinate system, you have not at all shown which one might be best in order to actually calculate some useful concrete result. I often make this point when talking to theoretical physicists who have climbed onto the band-wagon of coordinate-free physics. "How convenient!" I sometimes say, in sarcastic tone. On the other hand, I myself have published papers with results obtained by using unitary equivalence to go from an intractable physical situation to another more convenient mathematical coordinate system.

My next lecture was given at the private home and computational laboratory of the incredible Professor Kunio Kuwahara, who owned a city block in an expensive suburb of Tokyo. Kuwahara, a large man, was a kind of god to experts in computational aerodynamics all over the world. They would flood to his summer schools. He owned his own Fujitsu computers and employed several engineers there at his home-laboratory. We spent a wonderful day there and then settled into his expansive living room for an exquisite dinner prepared by his wife. Taking me to a wine shelf, he showed me a very expensive old Rothschild, and said we could open that one if that was my wish. I told him that would be fine – but did he happen to have something cheaper on the shelf that would be as good? He beamed, and asked me to choose any other bottle there. I chose a Haut Brion; Professor Kuwahara was delighted, for that is indeed a fine wine. Several of us played the piano and I still remember his youngest son, Tabito, meaning traveler. I have not seen Professor Kuwahara since; he and his family passed through Boulder some years later and looked for me, but I was out of town. I only learned during my 2010 lectures in China that Kunio Kuwahara had died in 2008.

I next gave two lectures at Waseda University, a wealthy private university important to Japanese life in a role not dissimilar to the Ivys in the United States. There I was hosted by the warm and friendly Professor Mitsuhara Ohtani, whose spoken English was almost flawless even though he traveled very little. I am continually struck by how some of us communicate and learn most easily by visualization, (where I put myself), and by how others just pick up and retain everything they hear. Discussing this with Ohtani, he smiled disarmingly and just said, "Well, I am a city boy, that's all I need." How could anyone not like this man?

From Tokyo we went down to the famous RIMS research institute in Kyoto where I gave two more lectures. We spent several mystical days exploring the Buddhist temples of that fantastic city.

* * *

I would return to Japan three years later for the XXIst Solvay Conference on Physics near Nara, the old capital of Japan. I gave another lecture for Professor Ohtani at Waseda and also a lecture at an IBM workshop on wavelet technologies. It was in discussions during this trip that I learned that Japanese mathematical and physics cultures, which developed rather independently from our Western Newtonian physics, look at physics as more of an art, a finding of patterns and harmony. We have for example our Conservation of Momentum Laws and other Laws of Physics, but we call them Laws because of our Christian culture – as if they are Laws of God! They are not, and more and more I am convinced that nothing is every exactly conserved, and everything is irreversible, all of the time, unless you can wait an essentially infinite time to see some things come back together again.

I had from the beginning of my studies in physics never felt comfortable with how my teachers would treat their equations as absolute truth. I would always see issues, asking for example, "Well, what boundary conditions are you assuming?" I would be told, "It doesn't matter," to which I might reply, "Well, is there anything else, anywhere, also interacting with your system?" and so on. So much of our science is simplification; granted, we must simplify in order to be able to calculate some concrete result, but it is simplification nonetheless, and therefore not a Law, and surely not a Law of God.

Meeting an Old Mentor at Solvay

There is no question that my affiliation with the Solvay Institute in Brussels, beginning in 1980 and ending only with Professor Prigogine's death in 2003, connected me to many luminaries in the field of theoretical physicists. However, when Ioannis Antoniou, who was then deputy director of the Solvay Institute, slyly called me to come down to his office "to meet an old friend" in the summer of 1996, I would never have guessed that it would be my old mentor Günter Lumer. Lumer lived in Brussels and taught at Mons, and was generally the godfather of the European consortium of mathematicians doing research on dynamical systems. Naturally I was absolutely delighted to see Lumer, who always beamed at me, and generally at everyone.

Lumer died in 2005. I made a quick trip to Mons Belgium and Valenciennes in Northern France in 2006 for a memorial conference for him. At the first evening session, there were two prearranged speeches about Lumer. Both struck me as rather dull. Then the Chair asked if anyone else had anything they wished to say. No one budged, so my hand shot up. The Chair was delighted and said, "Karl, come on down and give us your thoughts." Somewhat spontaneously, I recounted the following true story, mixing English and French as seemed appropriate to facilitate the message.

I had submitted a paper to the Pacific Journal of Mathematics in 1967, very early in my career, and soon after I received an unsolicited letter from a Professor Lumer. He invited me to come up to Strasbourg from Geneva, where I was on a second postdoctoral, to give a lecture and "talk about my recent result." In Strasbourg, Lumer immediately identified himself as a referee for my paper, and proposed that we write a joint paper generalizing my ideas. I was thrilled to have the chance to work with such an accomplished mathematician, and after several months, I sent him a first draft. When I arrived back in Minnesota, a letter from Lumer awaited me. It read, "Good work, Karl...but I think one can do more...why don't you try the following?" I did, working several months on it. I sent the enlarged paper to Lumer. A few months passed and I took a position at the University of Colorado, where I found a letter from Lumer awaiting me. "Good work, Karl...but I think one can do more...why don't you try the following?" I did, and after about a year, I sent an improved paper back to Lumer. After a few months, it came back to me with, "Good work, Karl, but I think one can do more...." to which I promptly shot back a reply:

"Yes. One can do more. But who is ONE?" Very quickly I received a response from Lumer: "Okay, I think we can publish it now." The audience, who knew Lumer's desire to always generalize a result a bit further, loved the story. Afterward several came forth to tell me that was why Lumer published relatively few joint papers. His would-be coauthors usually just quit from exhaustion.

Diving the Barrier Reef

The telephone rang in the motel room at the Ayer's Rock resort in the Australian Outback where my son Garth and I were leisurely packing to catch a direct flight to Townsville, and there join a dive boat out to the Great Barrier Reef. It was the manager of the motel: "Quantas has cancelled your flight." "What? How can they..." I started to say, but the manager interrupted me. "They are holding another plane on the runway, waiting for you two. They will fly you to Sydney, then up to Brisbane, then up to Townsville. Can you be ready in five minutes? I will drive you out to the runway!"

Fortunately my son and I were traveling light, as I always do. We finally arrived late at night in Townsville and the next evening headed out on one of Mike Ball's dive boats for a week of unlimited scuba diving on the Barrier Reef. Anyone who is a certified scuba diver needs no description here of the beauty one finds below the ocean's bland surface.

I had come down to Auckland, New Zealand, to speak at a conference on unconventional methods of computation – for example, using quantum or biological computers. My son flew down to join me at the end of the conference and we went to South Island and walked the Milford Track there, then returning to Auckland and over to Melbourne in Australia. With a free day there, we found that by chance the Australian Open tennis tournament was taking place in its early rounds, so we went out and enjoyed watching, among others, a Swede named Magnus Gustafson play a match. It was striking how hard and low these professionals could hit the ball. Thoroughly sunburned, we had then flown up to Alice Springs and Ayer's Rock. After the dive week, my son would return to his job in Boulder and I would close out the trip with a lecture at an applied mathematics conference in Coolangata and at a conference on neural networks in Brisbane. I was on sabbatical this spring term, thus able to accept some speaking invitations that normally I would decline, because I don't like making a habit of missing a week of classes when I am teaching a course.

This situation is peculiar to the United States, and especially to my institution. In many countries of the world, if a professor is noted enough and fortunate enough to be invited to speak in a major way at an international scientific conference, he will have regular assistants to turn over his course teaching duties to in his absence. Even secondary school teachers in the United States have a cadre of substitute teachers to call upon if they're sick or need to give an invited presentation. But not so at our universities. Moreover I often must pay most of my travel expenses, beyond those provided by the conference organizers.

Another incongruity is the situation with sick-leave. I earn sick-leave at the rate of one week a semester. Having now 40 continuous years at the University of Colorado, and having never taken any sick days, I have built up 80 weeks of sick-leave, or 400 days. Teaching a Monday-Wednesday-Friday schedule for a 16-week semester would mean I would use about 50 days of sick-leave if I couldn't meet my teaching obligations for the whole semester. This would imply that I could be sick with pay for four years now. If you ask me to be sick all five days Monday–Friday, I could be sick with pay for two and one-half years.

But the state budgets absolutely no funds at all for sick leave. When one of our colleagues falls into failing health or needs a serious surgery and recovery, he and his courses are simply not replaced in the budget. Each department must just absorb the additional load.

However, we do get our paid one-semester sabbatical leave each seven years. So I can go give international invited addresses and go scuba diving with my son. The cup may not be full, but neither is it empty.

Three Visas to Iran

That spring semester also found me invited to give the Keynote lectures at a major national conference in Isfahan, Iran, in May 1998. This was a rather dramatic, and I must say, tenuous development. But as I was not teaching that semester, I gave them a tentative, "yes," and waited to see what would happen. What would happen was that two times the organizers were able to secure a permission from the Iranian government for me to accept, twice that permission was cancelled, but in a third, very last-minute effort, they managed to get a visa to me. And with only minutes to spare, I boarded an airplane out of Denver with just 45 minutes in Chicago's O'Hare air terminal to transfer to an international flight over to Frankfurt.

The conference was the Second Iran National Seminar on Dynamical Systems, May 1–3, 1998. About 200 Iranian mathematicians and engineers would be flown in to Isfahan, which is in the center of Iran. The preceding year the Keynote lectures were given by a noted Russian mathematician. Spearheading the movement to have me be the annual lecturer there in 1998 was a chemical engineer who had taken a number of mathematics courses from me in Boulder about ten years earlier. The Mathematics Department there at the Isfahan University of Technology was in full support. I accepted the honor and invitation on November 25, 1997, five full months ahead of the conference dates. Virtually all of my expenses would be paid. The visa application process then began. I was to send all needed information and documents to the conference organizers in Isfahan. Iranian government regulations required that the visa request emanate from them.

But nothing happened.

Fortunately, in the coming months, email to Iran opened up. Khatami, a liberal, had been elected president by a 2/3 majority. There was optimism in the air. I had directed two Iranian students to their doctorates in Boulder and I had some feeling for the culture and personality of Iran. I was excited at the prospect of going there.

My only previous visit to any Middle Eastern country had been a lecture given and a few days in Istanbul, Turkey in 1995.

Then the organizers were informed that a visa had been approved for my visit – but then cancelled at a higher governmental level.

They renewed their efforts. A problem was that there was no U.S. Embassy or Consulate in Iran. The Embassy of Switzerland served to protect U.S. interests in Iran. Moreover there was no Iranian Embassy in the United States. Their interests had to be handled by a small office within the Embassy of Pakistan in Washington.

Through these contacts with the Minister of Higher Education in Tehran, my would-be hosts again obtained permission for my visa. But again it was blocked at a higher level. And I had no plane ticket, which was to be provided via Iran Airlines working with Lufthansa. The plane ticket could not be issued until I obtained the visa.

I was advised to pay for a special Visa Service which for a considerable fee would try to hand-walk my visa application through the Iranian Interests section within the Pakistan Embassy, thereby finessing the blocks being thrown up in Tehran. To do so, I must mail away my passport to them. With considerable trepidation, I did so. Meanwhile the organizers somehow obtained an official plane ticket which had been telexed to the Lufthansa office in New York. It was now April 18. I needed to fly on Tuesday, April 28 at the latest. Under pressure from Tehran, the Iranian Air portion from

Frankfurt was blocked, using the technicality that on my return flight from Tehran to the United States, I did not have a required four-hour interval in Frankfurt. I then met this by having Lufthansa schedule a 22-hour layover at a hotel in Frankfurt. But then my special Visa Service called and said they had done their best and had delivered my passport to "someone" at the Pakistan Embassy, but had obtained no reply and could do no more. By now it was Friday, April 24, an Iranian Sabbath in which all official offices are closed. Khatami's liberalization had now progressed to open phone lines and I was able to call my colleagues in Isfahan and inform them: still no visa.

Monday morning on April 27 at 7 a.m., I received a phone call from a Mr. Sinai in the Iranian Interests Office. I can still hear his voice, "Professor Gustafson? I am here to help you." Within one hour he called me back saying he had my passport and a visa. He would Fedex them to me. Those would arrive Tuesday morning, the flight day, at my residence in Boulder. I called Lufthansa, and they routed me through Chicago, arrival 3:46 p.m., departure 4:30 p.m. to Frankfurt. My son was waiting to rush me to Denver to begin the voyage. I made it. My chemical engineering colleague and a contingent of five Isfahan mathematicians met me in Tehran and we all flew down to Isfahan together. I had a wonderful 12 days in Iran.

Professor Inviteé a Bordeaux

Those interested in computational fluid dynamics obtained a one month professorial stipend for me in May–June 2000, so I might participate in a workshop there and give a couple of lectures. There are two amusing vignettes attached to that visit.

The first occurred when a former girlfriend from Brussels contacted me on my first day in Bordeaux. She informed me that she was flying down the next morning to see me, and maybe we could take a weeklong road trip together. In a state of consternation, I rushed to the office of my host, Charles Henri Bruneau, and told him the situation. After all, I was on salary...he interrupted me and in gallant French fashion, even with admiration, said, "Mais, il faut l'accepter, Monsieur, bien sur, bonne voyage...et vasi a ton aise!" (Go, and take your time.) Dominique and I drove up to the chateaus along the Loire. Also we were able to locate a small vineyard in St. Emilion and arrange a private tasting of one of my favorite wines, Chauteau Chauvin.

My month's salary was deposited in a French bank and shortly before my departure I went to their small local office and was assured by a middle aged gentleman there that a check would be mailed to me at my address in the United States. But none came. Unluckily for him, during my stay in Bordeaux, I had been invited to another workshop in Luminy, France, in August. While there I made my way to an office of the same bank in Marseille where I explained the problem. Greatly embarrassed, they checked some records, and issued me a check.

It definitely helps to be able to speak the native tongue!

Return to the Land of My Grandfathers

In 2002 I gave my first lectures in Sweden. In fact I was able to create a month-long lecture tour with seven lectures covering four Nordic countries. In some sense, I had decided to enjoy this landmark trip as that of a triumphant return of a grandson. Another ingredient was acceptance of advice from friends and colleagues: "Look Karl, you are over 65, still producing prolifically, paying part of the expenses of these international speaking invitations – you should view these trips as vacations!"

The first stop was at Roskilde University in Copenhagen, Denmark, hosted by Professor Boose-Bavnbek, whom I had never met. My lecture was about a mathematics professor Victor Gustaf Robin, who lived in France 1855–1897 and was denied tenure, and thereupon burned all his mathematical materials and died. Boose-Bavnbek is a German Marxist who was denied positions in Germany. But he did not seem to perceive the analogy. Late in August, he organized a conference on Mathematics and War in Sweden. I was invited and could not go, but was asked to send a paper for the Proceedings anyway. Probably they expected me to talk in some way about the mathematics related to my work for the military in the 1960s. Instead I wrote a fine piece entitled: *Two Farms*. It depicts how one farmer has unlimited children who eventually must over-run the farm of his neighbor, who had conscientiously limited the number of his children. The inevitable result is war. The editors refused to publish it. Why do people get so trapped within their own agendas that they can no longer see the larger perspectives?

From Copenhagen I joined my friend Professor Maurice de Gosson in Karlskrona, Sweden, and from there took the train to Sweden's west coast to seek the roots of my grandfather, Carl Gustafson. When I stepped off the little train at the

La Holm stop on this first leg of my trace of my Swedish ancestry, it was a bright, sunny day and not a soul was there. A single taxi arrived about 15 minutes later and the female driver was delighted to see me, to see anyone, standing there. She dropped me in front of the La Holm Stadshotell, where I was to spend the night. The door was locked. Again about 15 minutes passed and suddenly the door opened and the cook was very surprised to see me. He explained that the owners, having no guests at all, had decided to go on vacation to Italy. But have no fear, he said, I could have a room for the night. The next morning he and his kitchen help served me a full Swedish breakfast. I was the only one, besides them, in the elegant dining hall overlooking the river. The Stadshotell is a remarkable old building which contains a classic theatre/concert hall of one hundard seats. I have no doubt that my grandfather Carl Gustafson and his family stayed and dined there in the three years, 1897–1890, when they returned to La Holm from America.

My second lecture was in Trondheim, Norway, where my colleague Olaf Njasted managed to put me in the same hotel as the news media covering the storybook marriage of the royal princess Märtha Louise and the playboy Ari Behn. I saw it all: wonderful.

Then I flew to Uppsala, Sweden, where I gave a lecture to the Applied Mathematics Department and another lecture to the chemists at the Angstrom Laboratory. At the Applied Mathematics Department was Bertil Gustafsson, who had taken a functional analysis course from me in Boulder 30 years earlier. Next I flew down to Växjö in southern Sweden for a Foundations of Physics conference organized by Andrei Khrennikov, and also gave a mathematics lecture in Karlskrona at a conference on partial differential equations organized by Maurice de Gosson.

Finally I flew to Helsinki and bussed up to Tampere, Finland, to lecture for Simo Puntanen to statisticians there. The four-nation Nordic tour was over.

But how did it come about in the first place? One theme in this book is the remarkable way our human trajectories connect, often by chance, to carry us into our future. An appropriate but *ad hoc* starting point may be taken as the 1999 Les Treilles conference in France, to which my Greek physicist colleague Ioannis Antoniou had invited the Italian probabilist Luigi Accardi. Accardi in turn invited me to speak at his conference in Levico Terme in January 2001. There I met the Russian Andrei Khrennikov, who was invited to lecture in Denver in Fall 2001. Visiting us in Boulder that fall semester was the French-Austrian-Swedish Maurice de Gosson, who I had not known or invited, but whose seminar I occasionally attended. Andrei wanted to come up to Boulder to lecture and discuss physics with me and the three of us had dinner where Khrennikov proposed to Maurice that they both invite me to speak at their conferences in Sweden in 2002. Andrei had lots of conference money, Maurice only a little, but he gulped and agreed – later we became good friends. I knew the Norwegian Olav Njåstad from a semester he spent in Boulder twenty years earlier, and from a continued fractions conference my nephew Dr. Philip Gustafson had organized in Grand Junction, Colorado, in the summer of 2001. Norway back to Uppsala, Sweden, was facilitated by Owe Axelsson, who was a professor in the Netherlands. I had first met him in 1996 when he invited me to speak at his conference on numerical PDE in Nijmegen.

The connection to Finland and Simo Puntanen, whom I had never met, came about only because he knew my work from a paper on mathematical statistics which he refereed in 2000.

If you are a world-class research mathematician, you have entered an international community of scholars. Through that community, I have seen the world.

A Visit to Trinity

I had risen early and was driving south from Alamosa. It was October 13, 2004, and I was alone, on my way to Albuquerque for a regional AIP/APS/SPS/ΣΠΣ physics conference. After being student president of the Colorado chapter of the physics honorary ΣΠΣ in 1957, I had been inactive for 47 years. But a nice letter had come, inviting me to the meeting, and offering a special one day private ΣΠΣ tour to Trinity, the site of the world's first atomic bomb. My thoughts had gone to Stan Ulam, the famous mathematician who had hired me at Colorado, and who played a key role in the Los Alamos work on the atomic bomb. Stan had died in Santa Fe 20 years earlier, in 1984. I was also reminded of my mountaineering friend Bill Bueler, who had died from a fast-acting brain tumor only 6 months earlier, in April 2004. The last time I had seen Bill had been at the Great Sand Dunes north of Alamosa in August 2003. We had climbed some 13,000ers and then Bill said he had always wanted to make a grand traverse of the dunes. So we had walked north beside them for about six miles, climbed a bit higher up a dry slope into clumps of aspen trees until we were poised over the center of the dunes, and then headed straight south into them. One enters a land of tiny meadows followed by climbs up dunes again.

So why not drive the back roads down to the dunes, spend an afternoon reverie there, overnight in Alamosa, drive down to Santa Fe, and then on to the physics meeting in Albuquerque and the Trinity site? It would be a sentimental journey for Stan and Bill. The afternoon in the dunes had been spellbinding, sunlight plentiful, people sparse due to lateness of season, only a light wind. I sat atop the highest dune until sunset, enjoying that certain serene quality that aloneness brings in a place of such isolated natural beauty.

But overnight the weather had changed dramatically. A heavy rain had already changed to a strange sleet that remained coated to the road by the time I got to Antonito, Colorado. I had been looking forward to a quiet hundred miles thereafter through the high desert forest down to Espanola. The visibility worsened and I fell in behind a small truck so that I could follow its tracks in the developing snowstorm. Then the road forked and the truck continued right toward Chama, New Mexico. I almost missed the left turn. I was alone in a high desert blizzard. Soon I had to downshift and punch in the four wheel drive. Was it becoming more uphill? Or was it the frozen slush building up on the road? I could not tell. I downshifted again and was barely doing 10 miles per hour. I continued on, eyes straining to try to see the edge of the narrow road.

Suddenly I came to a gas station and diner at a small crossroads. I pulled off and got out. A good three inches. of frozen slush crunched under my feet. I was at Tres Piedras. The diner was empty. The station attendant told me I better hurry before the road was closed. About ten miles further south, the road would start dropping to a lower elevation, where the blizzard should lessen. Immediately I resumed the tense drive through the high piñon pine forest which I could not see. The road narrowed and went up and down and around curves. Suddenly at the bottom of a draw as I crept around a curve I was no longer alone. Out of the snowstorm came a beautiful column of about fifteen mule deer, walking calmly. Their coats looked very healthy, and quite dark, darker than those I usually see in Colorado. I stopped and let them cross the road a few feet in front of me. They quickly bounded up the hill to my left and disappeared. A beautiful encounter. Slowly I continued on. Soon I was losing altitude and was only in rain. I was on my way to Trinity.

Notes

My paper with Peter Rejto on the mathematical foundations explaining the atomic number limit 118 was, *Some essentially self-adjoint Dirac operators with spherically symmetric potentials*, Israel J. Math 14 (1973).

The quote of Mme Schlumberger was taken from a Fondation des Treilles booklet of photographs and descriptions published by Offset-litho Jean Genoud S.A., Lausanne, Suisse, 1988.

My lectures in Japan in 1995 were published in the book: *Lectures on Computational Fluid Dynamics, Mathematical Physics, and Linear Algebra*, Kaigai Publishers, Tokyo, 1996, and later by World Scientific, Singapore, 1997. Takehisa Abe and Kunio Kuwahara were co-editors.

Takehisa Abe and I published the story of Victor Gustaf Robin in two installments in *Mathematical Intelligencer* 20 (1998).

The world's first atomic bomb was exploded on July 16, 1945, atop a 100-feet steel tower at the Trinity site, in what is now a restricted area within the White Sands Missile Range.

My book *Introduction to Partial Differential Equations and Hilbert Space Methods* was first published by John Wiley and Sons, New York, 1980. Its second edition in 1987 was expanded. That version was translated into Japanese and published as two volumes, *Applied Partial Differential Equations* 1,2 by Kaigai Publishers, Tokyo, 1991, 1992. The ill-fated third edition, which essentially vanished along with Mr. Roy in India, was published by International Journal Services, India, in 1993. The revised third edition was published by Dover Publications in 1999.

Fig. 1 Carl and Sofie Gustafson, author's paternal grandparents. ca. 1920

Fig. 2 Edwin and Jeanette (Anderson) Gustafson, author's parents. ca. 1940

Fig. 3 The author on an early exploration. ca. 1938

Fig. 4 The author and his brother Dick. ca. 1945

Fig. 5 Living in the trailer. First car, a 1933 Plymouth convertible. 1951

The World Opens 89

Fig. 6 On the Diamond on Long's Peak, 14,255 feet, during the period when it was officially closed to all climbing. 1952

Fig. 7 Working at the gold mining operation in Alaska. 1954

Fig. 8 The author upon graduation at the University of Colorado. 1958

Fig. 9 Baidyaneth Misra, Josef Jauch and the author, in the Mathematical Physics library at the University of Colorado. 1971

Fig. 10 Becoming a single parent. Amy (11) on left, Garth (2) right. 1974

Fig. 11 The author and his son Garth near Boulder. 1976

Fig. 12 Harlan Barton, Julie Inwood, Richard, the author. Roger's Pass, Colorado. March, 1985

The World Opens 93

Fig. 13 The author just before his fall on the North Face of Lone Eagle Peak, Colorado. July, 1985

Fig. 14 Wives: (1) Becky with author, and children Amy and Garth. 1973. (2) Rose. 1978. (3) Jill and author. 1986

Fig. 15 Pierre Furrer, the author, Norman Bazley, and his friend Elizabeth, enjoying wine and good times in Baden, Switzerland. 1989

Fig. 16 Ilya Prigogine and the author near the author's mountain cabin. 1992

Fig. 17 Some lovers/friends. (1) Julie. 1982. (2) Diane. 1985. (3) Kathy. 1992

Fig. 18 Tibet: (**a**) With Tibetan nomads. (**b**) Norman Nesbit, the author, William Bueler, atop the Jokhang Temple in Lhasa. 1994

Fig. 19 (**a**) The author and daughter Amy at her graduation from Boalt Hall Law School, Berkeley, California. 1990. (**b**) The author and son Garth on the Milford Track in New Zealand. 1998

Fig. 20 The author at age 63, reliving youthful climbs in the Ampitheater. 1998 (Photo courtesy of Laura Middleton Downing)

Fig. 21 Robert Richtmyer and the author, near Gardner, Colorado. 2002

Fig. 22 With David Mumford during lectures in Vietnam. Cham Temple of Knowledge, Da Nang. 2005 (Photo courtesy of Jenifer Mumford)

Fig. 23 Lecturing on Game Theory with the letter from John Nash. 2006 (Photo courtesy of Allison Skidmore)

The World Opens 101

Fig. 24 Roger Penrose and the author at Millsite Inn, Ward, Colorado. 2007 (Photo courtesy of Homer Ellis)

Fig. 25 Plenary Lecture at Shanghai Finance University, China. 2010 (Photo courtesy of Yonghui Liu)

8. Personas and Personalities

...and a penetrating theorem...

There's that story that when God created the Universe, and the Earth, and the United States of America, and the favored professions therein, he wanted to give them good things. To the medical doctors he gave high salaries, to the lawyers he gave fascinating cases, to the carpenters he gave fine lumber, and to the professors he gave low teaching loads. Then he thought, I need to keep things balanced here, good means nothing unless compared to bad. So to the medical doctors he gave 4 a.m. surgery schedules, to the lawyers he gave vicious and unreliable clients, to the carpenters he gave unemployment, and to the professors he gave...colleagues.

The inevitable skirmishes with colleagues that anyone encounters in the workplace tend to be rather intensely intellectual within professorial life. Such intellectual arguments can be incredibly frustrating, particularly in view of the fact there is no right or wrong, per se, just a clash of values. So when it comes to colleagues in academia, one definitely encounters the good, the bad and the ugly! The bad and ugly ones don't deserve mention in this account. Some of the good colleagues, both in Boulder and throughout the world, appear regularly through the pages of this book. Many good, fine colleagues will not appear.

In mathematics and science, the role of mentors is particularly important. Some mentors stand out because of a powerful influence they have had on our perceptions of our own roles to be played within this enterprise, and more largely within our lives. Here are some of my own notable mentors:

Larry Payne

Larry Payne (Lawrence E. Payne), my Ph.D. advisor at the University of Maryland, was a wonderful mentor to me. When I met him, I was at a critical juncture in my professional life, as I emerged from the electronic espionage work within the military that had occupied me from 1959–1963. At that time, I could have continued on with the top-secret work. Or, being a rare computing expert then, I could have had my pick of well-paid opportunities in the burgeoning computer industry. However, with the awareness of my ever-present desire to eventually return to Boulder, I had begun a Ph.D. program in Mathematics at Maryland. Successfully passing all three Qualifying Exams, I needed a Ph.D. topic and advisor. When Larry offered me an NSF research assistantship under which I would not have to teach, immediately I sensed a warm, caring Ph.D. advisor with whom I would really like to work. And thus I entered into the field of partial differential equations.

Patiently Larry outlined several potential research problems within this vast domain, with all of its mathematical richness and wealth of scientific applications. In retrospect I was well suited to this field from my background in physics and engineering, much of which is modeled in terms of partial differential equations. I took two courses from Larry, one in fluid dynamics, the other in his specialty of Isoperimetric Inequalities. An example of the latter is one known already to the Greeks: Of all closed curves of fixed length in the plane, the curve which is an exact circle will enclose the largest area.

Larry Payne is short, stocky, and looks at you with his shirtsleeves rolled up as if he were an Iowa farmer going out to pitch some hay (or even shovel some manure). Usually he wears the slight fixed smile of one who is actively engaged with what he's doing. I would characterize his persona as: no nonsense, let's do it! He is still actively publishing new results in his favorite topic of inequalities at age 87, as I write these lines. Everyone likes Larry Payne.

Professor Gaetano Fichera, an Italian mathematician with whom I spent 6 months of my postdoctoral in Rome in 1966, was the exact opposite of Larry Payne. Dapper and always immaculately dressed in coat and tie and fine shoes, he ordered around his subordinates at his Instituto Matematico as if he were a dictator. He treated his lieutenant Professor De Vito like a dog, but De Vito never took it personally. Fichera liked elegant mathematics using a lot of "heavy machinery." One day he came by my office as I was trying to improve some inequalities for fluid dynamics, which Larry had already improved over those of the famous Russian mathematician Olga Ladyzhenskaya. After a little mathematical talk, Fichera looked at me and said: "You know, Gustafson, I like Payne. He's a nice fellow. But all he ever uses is Schwarz's Inequality." I replied, "Maybe that all he needs."

Tosio Kato

During my postdoctoral year 1965–1966, two great books were published by the Japanese mathematicians Kosaku Yosida and Tosio Kato. Yosida resided in Tokyo, while Kato had moved from Japan to Berkeley (where he was a math professor at the University of California, Berkeley). Yosida's book *Functional Analysis* went on to many editions and set a standard for that subject as an abstract treatment of differential equations and other important topics in mathematical analysis. I would use it today were I asked to teach such a course. Kato's book *Perturbation Theory for Linear Operators* is similar but more specialized, in particular toward the operator theory needed to make quantum mechanics mathematically rigorous. In the 1940s, Kato shared independently with Franz Rellich of Germany the first proofs which put the famed Schrödinger equations (which describe quantum mechanics and the chemical elements) on a firm, rigorous mathematical foundation. I more or less fell in love with Kato's book and its sparse, perfected style. As my Hungarian-American colleague Peter Rejto, with whom I once wrote a joint paper, liked to say: "Kato is God." I improved one of Kato's results and am honored to be cited in a later edition of his classic book.

In 1975 I proposed Kato as the annual lecturer for our Mathematics Department. He won the selection process, thus putting me in charge of arrangements for his one-week visit in Boulder. Previously I had spoken with him only fleetingly at a couple of meetings, and knew little about him personally. When I went to pick him up at the airport, there was this slight, totally reserved, now older man, and our conversation was limited to extreme terseness on the hour-long drive from Denver to Boulder. When we stopped at my house briefly before I was to take him to the Faculty Club, he informed me and my wife Rose that he suffered from heart problems and would need a totally cholesterol-free diet for the week. We did our best, either at my house where Rose prepared special meals, or at a few select vegetarian restaurants. His three lectures were, of course, impeccably prepared and delivered in an efficient and spare style. Professor Kato was the epitome of the inscrutable Asian personality.

Rose died in March of 1980, and ten months later I managed to get myself out to the American Mathematical Society annual meeting in San Francisco in January, 1981. Some months before, I had sent Kato a complimentary copy of my new PDE book. After I arrived in San Francisco, I received a phone call from Kato at my hotel, saying he would not attend the meeting, but he would like to invite me to lunch nearby. I was delighted but surprised, as he was one of the most reserved individuals I had ever known. I met him at the restaurant and he thanked me for my book. He then mentioned a colleague, a German mathematician, who had told him I would be at the San Francisco meeting. He was very sorry to have heard that Rose died. He said he remembered her as one of the warmest, finest women he had ever known. He told me how much he had enjoyed the whole week spent in Boulder, due in large part to her preparing meals and her caring nature toward him. He was very sorry for me, he said; but surely I would go on and do some more fine mathematics? How were my children and Rose's children doing? He was confident in me to take care of them. Lunch was quickly over and we bid each other adieu. Kato's persona I would have to describe as: ultimate reserve. But he went out of his way to show me rare compassion, which I appreciated tremendously.

Kato passed away on Oct. 2, 1999, at age 82. Although he had retired from Berkeley in 1988, he remained active in the field of applied mathematics until his death.

Josef Maria Jauch

From the time I first attended his petite seminar at his Institut de physique theorique in Geneva in 1967, until his untimely death in 1974, I was much influenced by Professor Jauch, and we had many, many deep discussions about the foundations and meanings of quantum mechanics. Jauch was typically Swiss looking, with a demeanor not unlike that of Albert Einstein, a ruffled look, and large soulful eyes. However, he did not play the role of genius; rather, I came to view his persona as: dreamer.

Jauch wrote a compact little book called *Foundations of Quantum Mechanics*, which he wryly described in the first sentence of the Preface as "an advanced text on elementary quantum mechanics." Jauch was truly a philosopher of physics and had built up his school in Geneva to rival the famous school of Einstein, Pauli, and others in Zurich. Many young mathematical physicists who were my peers had spent a year or two with Jauch: Kalyan Sinha, Baidyaneth Misra, Jean Pierre Antoine, Vaughn Jones, Ann-Marie Bertier, and Larry Horowitz, among others. When in 1970 I could invite one personage to Boulder to help kick off our new Mathematical Physics Ph.D. program, I invited Professor Jauch.

He was divorced and I had the impression that he was living as a lonely scholar, although his son Karl also lived in Geneva. Once, he took me and my wife Becky sailing on his boat on Lake Geneva. Becky was a looker with classic legs, slim hips, and a full bust. She was wearing a "Daisy-Mae" outfit that day, with tight Levi shorts. Professor Jauch could not help looking lustily at her, and exclaimed, "Becky, you look like Eve! But you need an Adam!" She smiled, a bit uncomfortably, at him, looked at me, and replied, "But I have my Adam." This was not completely true because our marriage had cooled off after ten years. But she had detected a longing, even an intimation of a possibility, from him, and she did not welcome it.

About a year later Professor Jauch met an elegant French woman who had two teenage sons. He decided to marry her in Boulder, and I and the Chair of the Physics Department, Wesley Brittin, made the arrangements. The marriage was a disaster and lasted less than a year. The next time I saw Professor Jauch in Geneva he told me, "She ruined me, Karl." He was driving a motorbike because he could no longer afford a car.

Jauch's death at age 59 in Geneva in August, 1974, was shrouded in mystery, but from close colleagues there I learned later what is reputed to be the actual story. Jauch had taken a young female postdoctoral student over to his son's apartment in Le Lignon, a high-rise apartment building complex. I know that building because my family and I once stayed in that apartment complex for two weeks while hunting for a year's rental in Geneva. The elevators worked only in limited fashion and slowly at best. I was told that Jauch had bedded the young woman and died of a heart attack. To save a scandal, his young colleagues, some of whom I knew, had to bring him down that slow elevator at night to take him to his home. I can believe this story. But it does not diminish my respect for one of the deepest thinkers about quantum mechanics I have ever known, and a man of unusually generous nature.

After his death, his lieutenants maintained his school at a high quality even until today. But I think Jauch was the soul of that school.

Ilya Prigogine

I first met Ilya Prigogine in 1978 in Brussels when I was visiting my colleague Baidyaneth Misra there at Prigogine's Solvay Institute for Physics. Prigogine had just been awarded the Nobel Prize in Chemistry in 1977. Shortly after in 1979, I and my colleague R. K. Goodrich solved a mathematical problem that had been blocking Professor Prigogine and his coworkers at his institutes in Brussels and Texas. Prigogine was in Texas and excitedly called me to see if he could come up to Boulder to meet us, and, by the way, he would be willing to give a lecture. Frantic, I called acquaintances in the Chemistry Department for some money and some help. In a matter of days they arranged for Prigogine to give the annual Edward Condon Memorial Lecture in Chemistry here. These events initiated more than 20 years of close collaboration for me with Professor Prigogine and his team at the Solvay Institute in Brussels, in which I became an Honorary Member. We had many conversations together and became good friends.

Prigogine's persona became that of: visionary. After the Nobel award in 1977, he decided to speak and publish openly and widely about his visions of physics, science, and society. This made him a hero to nonscientists but caused some resentment among scientists. Early in his life Prigogine had been interested in philosophy and ethics, and not just science.

Prigogine made clear to me on many occasions that he opposed Einstein's view of basically deterministic laws of nature. On the other hand, he shared Einstein's position that fundamental physics cannot depend on the subjectivity of human observers. He wanted something objective, yet not deterministic. Based upon his conviction of an underlying stochasticity of the universe, he was always searching for some mathematics to explain the direction of time. Because of my association with Prigogine, I wrote some papers placing some of his intuitive conclusions into proper mathematical formulations.

Professor Prigogine did not conceive the future as determined by one universal theory to which everything could be reduced. With Bohr he believed one must give up the realism of classical physics. In an influential book with Isabelle Stengers, *Order Out of Chaos* (1984), he wrote, "The irreducible plurality of perspectives on the same reality expresses the impossibility of a divine point of view from which the whole of reality is visible." Hoping, like Einstein, that mathematics could provide conceptual breakthroughs in our understanding of nature, Prigogine became a patron of mathematical physics. I was lucky to have been his confidant. He liked very much to bounce ideas off me. Often we disagreed because I could not support his intuitive conclusions with mathematics.

Baidyaneth Misra and I teamed to direct the dissertation of a brilliant Greek physicist Ioannis Antoniou at the Solvay Institute in Brussels. Ioannis became Deputy Director there, my friend, and we wrote a number of scientific papers together. I was privileged to speak at the 21st and 22nd Solvay Conferences in Physics in Japan and Greece in 1998 and 2001. At the former I presented my new trigonometry for quantum probabilities. At the latter I sharpened the famous Bell inequalities of the hidden variable controversies into new trigonometric Bell equalities, from which I concluded that one cannot argue non-locality solely on the basis of violation of Bell's inequalities. This issue still divides physicists and only better experiments will finally settle it.

Professor Prigogine died in May of 2003 at age 86. The last time I saw him was in June, 2002, after I returned to Brussels from my Scandinavian lectures. Ioannis had installed me at Prigogine's desk in his palatial office at the Solvay Institute, a large room filled with exquisite archeological artifacts. Professor Prigogine had, however, unexpectedly returned from vacation, and I was surprised, even embarrassed and startled, when he and his lovely wife Marina walked in. He went over to the extra large table in the room and said, "Don't worry, Karl, you can keep on working there." But I didn't.

One day in Brussels in 1987, over lunch Ioannis and I pondered the question: Having already won the Nobel Prize, why was Ilya Prigogine still pushing so hard? At 70 years old, he directed the Solvay Institute at the ULB and also maintained his institute at the University of Texas. He was writing books on his visions for society, he enlisted numerous young collaborators to write joint technical papers and – as our many conversations with him evidenced – he was always searching very hard for some mathematical equation which could extend thermodynamics all the way down to the level of quantum mechanics.

For some reason, this question, a speculation on Prigogine's core motivations, played in my mind all afternoon. At dinner alone in Brussels, I had an insight: Why not try to answer it in Einstein's way, with a Gedanken Experiment? True, Einstein always used his Gedanken, or thought, experiments, to try to resolve issues in theoretical physics. The method was to set up a hypothetical but cutting edge imagined physical situation, and ask a tough question. Such Gedankin experiments were used by Einstein throughout his conceptual battles with Niels Bohr about the true nature of quantum mechanics.

Over continuing wine and still with Prigogine in mind, I scratched out the following prescription, which I just wrote as: The Theorem. You could as well call it the Algorithm, but I prefer the term Theorem because it states, or purports to state, a general truth.

Theorem

(Name), wants, above all, more than anything, (what)?
Corollary 1. In seeking (_), he/she will never get enough of it.
Corollary 2. In seeking (_____), he/she will go too far and will actually obtain less of it, thereby remaining frustrated.
Corollary 3. In seeking (_____), he/she will damage those around him/her.

The idea is to place the name of the individual in question in the first blank, and then force yourself to fill in the second blank with a single word. That single word is then placed into the blank in each of the three corollaries, as a check on, or further confirmation of, your choice of single word in the Theorem.

For best results, you must know the subject individual very well.

For Ilya Prigogine, I wrote, <u>Immortality</u>.

The next day at lunch with Ioannis, I presented him with a sheet of paper with my Theorem and three corollaries and all blanks. I told him to put Prigogine's name into the first blank. Then I told Ioannis, "Think hard now, take deep breaths...but fill in that second blank with one word." After a few minutes of silence between us and some breathing, looking around, clearing the mind...Ioannis wrote down, <u>Immortality</u>.

* * *

The corollaries need not always work or even be appropriate for the individual. But they can serve as a valuable check on your single key "want" word. You may or may not be able to apply the Theorem to someone you do not know well enough. Some days later, Ioannis and I tried my Theorem on ourselves. Ioannis wrote for me, <u>Joy</u>. I wrote for Ioannis, <u>Truth</u>. We both knew my colleague and close friend in Boulder, Kent Goodrich. For Kent, I wrote, <u>Security</u>. Ioannis independently wrote, <u>Stability</u>. Later Kent accepted those as valid answers. Kent will tell you that he is extremely risk-adverse. Kent wrote for me, <u>Challenge</u>. There need not be uniqueness of answers from different observers. The idea of the checking Corollary 1 is that the *want* of the Theorem is so strong that it is really a *need*. Corollary 2 may identify the *want* as in fact a *compulsion*. Corollary 3 in some sense identifies the *want* as so strong that the individual's single-minded drive toward it may ignore equally valid *wants* of his/her associates.

I decided then to try to apply my Theorem to all of humanity. <u>Humanity</u> wants, above all, more than anything, (_____). My answer was, <u>Meaning</u>. Ioannis, independently, wrote <u>Significance</u>. Go ahead and put your *wants* word into the blank. Then the Corollaries can be interesting for you. Not only may they affirm your choice, but they can raise interesting issues. For example, for me, do religions at times go too far and actually do damage to others, e.g., the nonbelievers, the infidels? Tribalism also provides <u>Meaning</u>, but may require attacking other tribes to maintain or increase hunting territory. In such ways, the Theorem can become a sociological Gedanken experiment.

* * *

In the period of 1987–1992, the University of Colorado administration decided to split our Mathematics Department into two completely separate departments: (Pure) Mathematics and Applied Mathematics. There were both good and bad reasons for doing this. But being both pure and applied mathematicians, I and some others fought this divorce – and we lost. Somewhat in frustration, as I watched what was in my view the unfolding of a tragedy, I tried out my Theorem on some of the active participants. It did not work perfectly because I didn't know them well enough. However, I asked one or two trusted colleagues to also fill in the individual's name and consequent key driving *want*, as least as seen from their role in this crisis. Without giving names, here are some examples on which one or two or three of us essentially agreed.

A colleague who tried to straddle the issue, even play one side against another, had the *want*, Control. Ironically, his mathematical specialty was control theory, although that is probably irrelevant. As to the Corollaries, he quickly lost influence, and retired not long after. Another, our departmental chair from 1987 to 1990 and one who found himself uncomfortably caught in the middle of the crisis, had the key *want*, Approval. Of course, there was no approval to be had from either side and all three corollaries kicked in and he could not stop the split. A younger colleague who was one of the key proponents of getting a separate Applied Mathematics unit in place, and who it seems did some rather dastardly acts to keep it moving along, was accorded by three of us, the core *want*, Respect. He had been hostile to the Mathematics Department from the day he arrived, probably due to some innate feeling of inferiority. Indeed, he benefited from the split, and obtained much more influence, hence respect, within the new separate applied math unit.

So I think my penetrating Theorem can even be useful in helping to understand your enemy. But one must be cautious with it. Politicians will tell you to never argue the motives of your opponents, always keep it at the level of factual disagreement – or I might add, keep it at the rhetorical level. In other words, don't make it personal. On the other hand, my Theorem indeed looks at core, even soul, motivation. This can aid you in understanding, even helping, either friend or foe. So, not being a politician, I would even argue for the radical proposition that our civilization would benefit by demanding that motivations be made open, apparent, and admitted.

Recently I ran my Theorem by a colleague in our Department of Psychology and asked him what related theories came to mind. He told me of Attribution Theory. An example he mentioned: Take a TV evangelist, turn off the sound, look at the face, and decide whether you see Sincerity or Insincerity. Another example is that of laying blame, whether it is deserved, or not. My Theorem has a deeper motivation: trying to understand those whom you love.

* * *

I have never before applied my Theorem to my beloved mentors Larry Payne, Tosio Kato, and Josef Jauch. So I do so now. Larry Payne, whom I have known well for more than 45 years, wants, above all, more than anything else: Work. Tosio Kato, whom I never knew closely, but whose book I studied intimately, wanted, more than anything else: Perfection. Josef Jauch, wanted, above all, more than anything else: Love.

Finally, if you do not find a single word rising into your consciousness, it is helpful to accept a phrase. For example, Larry Payne wants to Work with others. One may then condense this phrase into Collaborate. He greatly enjoys writing his papers jointly with fellow enthusiasts. Tosio Kato also wrote joint papers but because he wanted to set the highest standards in mathematical writing, most of his papers were sole-authored. Josef Jauch loved writing papers and books with his junior colleagues such as Piron, Amrein, Sinha, and others in his Geneva school of quantum mechanics. A competitor once tried to deprecate the Geneva group by creating and spreading the anecdote that, "Piron gets the idea, tells it to Jauch, who then gives it to Amrein to do." When he shared this with me, I replied, "Well, they get more interesting results than you do!" Jauch also loved associating with his many postdoctoral visitors. Taking into account his personal life as well, a somewhat refined answer to my Theorem's application to Josef Jauch could be: Affection.

These four mentors who were so important to me displayed quite distinct personalities. Larry Payne is naturally unassuming and friendly, no doubt an extrovert. Tosio Kato was an introvert but with a deep kindness, conditioned by his perilous youthful years under bombing in Japan in World War II. Josef Jauch came from the traditions of a well-established Swiss family which probably afforded him the leisure to become a dreamer. Ilya Prigogine could appear very pompous on the surface, but was absolutely open and honest in his discussions with those he trusted. I think the public should see scientists, even mathematicians, as often possessing interesting, and in some cases even fascinating, personalities. We are not robots just because we must often exercise our left brains. In fact I would assert that the more successful and creative scientists and mathematicians whom I have known throughout the world have very active right brains, and full and quite interesting personalities.

Although there is no doubt some relation of my Theorem on understanding core overwhelming motivation of an individual, to understanding that person's personality, I think that most of us would prefer to keep internal motivation distinguished from personality. Obviating modesty, let me illustrate this point with myself. Above, two colleagues who had known me, respectively and at that time, about ten and twenty years, accorded me an overwhelming want of Joy and Challenge. Recently, a close female acquaintance who has known me only one year but with whom I have shared much of my life story and many philosophical discussions, found me always searching for Beauty. Which of these is the better answer to my Theorem? I apparently do have a strong sense of Joie de vivre. Although that passes Corollary 1, Corollaries 2 and 3 do not confirm it as an overwhelming need. In my opinion, the same can be said about my always seeking Beauty. So whether I like it or not, Challenge best satisfies the Corollaries. Joie de vivre rests probably better as an aspect of my personality. In failing one or more Corollaries, perhaps one has nonetheless identified an attribute of personality.

Not liking the idea that I am so driven by Challenge, I rationalize by offering the possibility that the Theorem allows you to try to refine the verdict of your peers, as I did with Josef Jauch above. I do find Curiosity as an overwhelming want of mine which gets me into a lot of situations that lead to confirmation of all three Corollaries.

In like vein, my close female associate has exhibited in our discussions and in her meditative practices a continual search for spiritual Bliss. But the Corollaries do not entirely reinforce that outcome for the Theorem. Upon further reflection on her whole life, she found herself answering the Theorem with: Freedom. I can only partially concur, and note that Corollary 2 is only weakly confirmed. This brings out an essential point: The Theorem has most force if it is not answered by yourself, but rather, by those who have known, and perhaps loved you for a long time. Evidently I have not known her for a long enough time to answer the theorem for her.

Notes

More details about Professor Prigogine's scientific views and of my interactions with him may be found in two invited memoria: *Professor Ilya Prigogine: A Personal and Scientific Remembrance*, Mind and Matter **1** (2003), and *Microscopic Irreversibility*, Discrete Dynamics in Nature and Society **8** (2004).

9. Wives, Lovers, Friends

...and a great loss...

Anyone with a rich personal life has both bright and less bright episodes in their past. Some choose to lead a more sterile life in order to not risk emotional setbacks. Upon review, mine has not been a sterile life. But I was a late bloomer.

I had a crush on a cute "sweater girl" in high school but she did not care a bit about me. She was after the quarterback. A cheerleader liked me and in front seats and back seats of cars we engaged in petting and necking (but not the real thing) up on Flagstaff Mountain above Boulder. That was the custom in those days. But I was not taken with her; although now when we meet at class reunions, I wonder why not. Somehow my chemistry was not yet turned on. I managed one conquest of a nice girl, I suppose to see what it was like. That was also on Flagstaff Mountain. It was not too memorable; but afterward she did invite me to the prom.

College life found me immersed in engineering school combined with business school – and poor, with no money to take girls out. Instead of going to proms, I ran a small floral business out of the fraternity kitchen, so that others could have their carnations to give to girls they took to proms. However, in my final fourth and fifth university years, some of the girls at the Chi Omega sorority, where I worked as a waiter, would buy the pitcher of beer if we would go dancing with them. We did a lot of jitterbugging and rock at the Tulagi's nightclub on University Hill. Another favorite place was the Timber Tavern at Arapahoe and Folsom. However, I had no steady girlfriend.

After receiving my double Engineering-Business BS degrees in May, 1958, I decided to stay on as a waiter at the Chi Omega sorority to augment my small salary as an Instructor of Applied Mathematics. A stunning Irish-German brunette named Phyllis came out to summer school from Chevy Chase, Maryland, and took a liking to me. I took her dancing, and up the third Flatiron, a 700 ft. rock climb above Boulder. But I was too inexperienced to foresee what was coming. Through her girlfriend she let me know that she wanted to stay in Boulder for a few days after the end of summer school. I arranged a room at the home of the mother of one of my climbing friends.

By the first afternoon in the room up there, things became so hot that we just jumped into my car and drove four hours over to Aspen. We made love all night in my sleeping bag, soaked in a light rain, on the grass beside Maroon Lake, looking right up at the Maroon Bells. Two more days and nights full of lovemaking followed before I put her on a plane back to Washington, D.C. It was my first, and a very torrid, infatuation.

Thanksgiving brought unrelenting lovemaking in borrowed cars and any parking lot in Chicago, where she attended a private school. I remember driving to South Bend, Indiana, to see Notre Dame play USC. That was my first real exposure to Catholic culture. Christmas vacation took me to Palm Beach to join her family in their home there. Then we drove back to Colorado for a little skiing, but mostly for lovemaking along the way. When we were on the road and nightfall came, Phyllis liked to cuddle right beside me. She would gently tongue my right ear until I could not stand it and we would pull over and do something about it.

It was because of Phyllis that I took an internship in Washington, D.C., in the summer of 1959. But our incompatibilities (she was wealthy, I not; she was Catholic, I not) became clear, so by the end of the summer, our paths diverged. Meanwhile, the top-secret project I had fallen into with the government had actually become more compelling. They wanted me to stay. But first I would return to Boulder.

That fall I started dating others, among them Becky Emeis, who would become my first wife. When Becky walked through the student union in her first year at CU, all conversations would stop and the male coffee cups would rattle. She was simply a knockout. At one point she was asked to model for Playboy magazine, but she did not. We met on a double date, I cannot remember who I was with, but her beau was named Dusty. When she broke up with him soon after, I carefully asked him, "Are you guys really broke-up?" He assured me that she indeed had dispensed with him. It took him years to get over her. I quickly asked her out.

Later that year I was drafted into the Army. As I have recounted in Chap. 4, complicated arrangements between the Selective Service, the Army, and my Navy colleagues put me back into the top-secret espionage work in which I had become involved during my internship. Becky and I planned to get married and move to Washington, D.C. The required three-months Army Basic Training at Ft. Knox, Kentucky, was delayed so I could finish a preliminary and vital new espionage software project for the Navy. Inter-service rivalries! We were married in June, 1961, and Becky transferred to the University of Maryland, from which she would graduate in 1963. The plan had been for her to then work to support my going to graduate school. Instead, our daughter Amy was born in 1963.

I would characterize this first marriage as one that could have endured for a lifetime, but which steadily cooled off. An example: to look good in daytime, Becky would always put her hair up in curlers at night. That is not conducive to romance. Our marriage was not an especially warm one. However, we were a competent, stable couple for 13 years.

Our son Garth was born in 1972 in Geneva, Switzerland, where I was on sabbatical. Becky now faced a big loss of freedom with another small child to raise, while our daughter was then 9 years old. When we came back to Boulder,

I found myself staying home a lot to share in the childcare. I changed a lot of diapers. That was not the normal behavior in that era.

Becky had always dreamed of becoming a proficient ice skater and she started taking skating lessons. Soon she was absent for longer intervals. I also could tell she was not really interested in being with me physically. Probably I also contributed to the dying marital passion with my mathematical preoccupations. Sometimes a mathematician becomes totally preoccupied.

I must say that the way in which I finally reacted to my suspicions has elements of both rationality and humor. The rationality told me: wait until the semester has ended. The humor told me: take up ice skating. I hired a babysitter and would go over to the rink and practice skating backwards while gauging the chemistry between my wife and her 19-year-old instructor. I am a terrible ice skater. But the conclusion regarding my suspicions was clear. Just to be certain, I tracked them over to his apartment one afternoon. Later, I confronted Becky when she returned home. She denied having an affair. I then suggested that we have sex on the sofa and we moved in that direction; but suddenly she just began sobbing and uttered, "I can't." What happened next surprises me still. My barely contained anger exploded into fury and I literally pushed her out the front door. Naturally she protested, but I was as resolute as I was furious. "Look, if you are sleeping with someone else, you are not going to be sleeping here!" I shouted. I am not quick to anger nor raise my voice, and never have been. But at that moment, possibly for the first time in my life, I was overwhelmed by my emotions.

The decision on my part to become a single father in 1974 took a lot of courage. The children were ages 11 and 2 at the time. My resolve did waver for the first few weeks. But then I noticed no more of Becky's cigarette smoke at eye level, which I had endured uncomfortably for 13 years. And I noticed that suddenly I was no longer always on the defensive. I simply felt there was no going back. Becky had hired a lawyer, but suggested that the affair would pass in a few months. I chose to file for divorce, and the marriage ended without much ado. She and her young lover really liked each other and stayed together for four years.

I was promptly cast into the singles culture of the 1970s. It was a pretty wild culture. In July of that summer I first ventured out, on a climbing club trip, and ended up that night in someone's bed. I became involved in Democratic Party politics and by August I was the steady bed-mate of the female candidate for County Clerk. Then at a political rally I met Rose.

Rose was a secretary in the Astrophysics Department. I had a second office over there in connection with a Mathematical Physics Ph.D. program I helped run. During the last two (unhappy) years of my marriage I had noticed her, a slim blond. I liked her demeanor, her walk, and how she would not make eye contact. We never spoke.

The political rally was at the Hotel Boulderado and was partially for my friend Clela, who was running for County Clerk. Rose and I sat outside and talked for two hours. Clela sensed that something was up and came out several times. Then she would go back in to do the wild free-style dancing that she liked. By the end of the evening she was, figuratively speaking, history. I was very taken with Rose.

She quietly told me her phone number, which I quickly repeated to myself several times so I would not forget it.

Rose and I met for drinks at La Paz, a local Mexican restaurant, which was, as she put it, neutral ground. The next time as we embraced in her living room in that rundown rented house, she asked me what I wanted from her. We quickly became lovers. But she had three children there with her, and I had two children at home. One night I hired an overnight babysitter and spent the night at Rose's. She and her husband were separated and had no money. He was unemployed and would sometimes come over in the morning to watch the kids when Rose went to work. That morning he came too early, opened the bedroom door, and I heard an agonized, "Oh...." A few days later he told Rose he would take the children from her if she had any more sleepovers. In those days he probably would have succeeded.

Rose and I decided quite quickly to get married and move all five children in together at my house. We did. It was still 1974. Looking back I see our decision as very risky. But it was one of the best things I did in my life. We provided a home for five children, and for ourselves. The children were Christopher, 13, Kelley, 11, and Patrick, 8, from Rose; and Amy, 11, and Garth, 2, from me.

It is a real challenge to raise five children, especially including stepchildren. We would seat the five of them, with Christopher in charge, at the family room table, while Rose and I would eat dinner in the adjacent formal dining room. For Christmas Eve, we instituted the tradition of a cheese fondue with white wine. Everyone so enjoyed it that I have carried on that tradition for all these years with the kids and friends.

Rose freed me up. For example, she introduced me to marijuana. Not as a pastime, but to enhance a nice evening of eating and intimacy. A year passed and after the next Christmas came and went, I was invited to speak at a conference in Mexico City. I had hired a "governess" when I was first single to let me get out on a Saturday or Sunday, and Rose and I had continued with this governess, who was a real character. Her name was Marie, she was raised elegantly in New York City, was 65, and was no-nonsense with the children. Leaving Marie in charge, Rose came with me to Mexico City. After the conference, we climbed Ixtaccihuatl, a nearby 17,000-foot volcano. On the other side lies the town of Puebla and its suburb Cholula, where there is a huge pyramid. We felt like smoking a little pot so our guide readily supplied it to us. He said to not worry about the police. So Rose and I sat on a main street at a small restaurant and became entranced with the incredible dancing leaves on the trees across the way and with just about everything else. Broad daylight and we were there for hours. After we were back in the States, I really wondered about our judgment that day!

The title of this chapter is Wives, Lovers, Friends. Rose was all three to me, and fully, and I to her. My first wife Becky may have been all three at the beginning, but more wife, less lover, and less friend. Of course, you are hearing only one side of the story.

Tragedy struck in the fifth year of our marriage. Rose had passed her annual physical examination with flying colors and was the picture of healthful beauty.

But in the fall of 1979 she lost her appetite. I made her promise to get a diagnostic exam while I was away at conferences in Europe. When I came back the diagnosis was ulcers. But the medicine did not bring back her appetite. We put her in a hospital in Denver and the diagnosis became Crohn's disease. But the old surgeon there did not believe it. An exploratory operation was scheduled to take a look. A one-hour duration was expected. Two hours passed. Three. Four. Then the kindly old surgeon came out to the waiting room. He looked at me and asked, "Are you alone?" "Yes," I replied. Then, reminiscent of a scene from a movie, he said: "You'd better sit down." I did. He matter-of-factly told me that Rose was riddled with cancer. It had started exactly where the large and small intestines meet, he explained, a notoriously difficult place in which to detect a developing cancer.

The moment was utterly surreal. It was certainly the worst moment of my life. The room seemed to spin as I fought to comprehend the meaning of his words. I could not breathe and felt I was about to pass out. Somehow, I managed to drive back to Boulder alone that night, shaken to my core.

We had a wonderful last six months of her life. Rose endured three more surgeries. Between them she was strong and we could even run together. We had lots of time to discuss everything. Her son Patrick would remain with me and my son Garth. Christopher was in college, Amy was going to college, and Kelley was under the supervision of her father. Rose insisted I not cancel a small American Mathematical Society conference I had organized and was to co-chair in Boulder in March, 1980. The second day of that meeting she invited an intimate circle of my closer international friends over to our house for a small reception. She greatly enjoyed them, and they her, even though she was walking around trailing an I.V. stand. That night she died. She had been on her feet to the end.

During her illness Rose as usual exhibited her generous yet down-to-earth nature. One day she said to me, "Instead of those with cancer saying, 'Why me?' they should say, 'Why not me?'" Rose also carefully warned me one day, "Get ready Karl, it is always harder on those left behind." She never complained about herself, but one day she did sorrowfully lament, "But I am not ready to leave the ones I love!" I dedicated my first book to Rose with those words, to the woman who had always encouraged me with a, "Go for it, Karl!" when I would get up early to work on that book.

Rose had always maintained the family photo albums. It is just not in my nature to do scrapbooks. Many years later, I opened up the last one. She knew that eventually I would get around to it. There, in the last page, I found her final note to me, just a scrap of paper, on which she had written

I love you more than words can say.
Hold on for me! Rose

Arranged around it were three photos: the one of her she knew I liked best, the one of us she liked best, and one of us with my little son Garth, whom she had so loved, between us.

Again, I was a single parent. I had been invited to Bielefeld, Germany, for a month-long summer workshop on molecular dynamics. I took Garth and Patrick with me. Afterward, we motored down to Switzerland where I had friends. From there we took a slow train to Bari at the southern tip of Italy, where we succeeded in booking overnight deck-sleeping space on a decrepit ship called the Poseidon, heading to Greece. After some days in Athens we found a small island named Elafonisi at the bottom of Greece. We camped on the beach dunes there for a week in 100+ degree heat, and then summer cracked. I still use this phrase, the day that summer cracks, for that day in late summer when the Earth's heating up ends with a certain kind of thunderstorm. Our new team had formed. Patrick, 14, had already become a big help to me. In many ways, his gentle, caring personality resembles his mother's.

I did not see any women for more than a year. Then Norman Bazley came to give an invited lecture. With Norman's great outgoing personality, Patrick and I decided we should have a large party at our house. We did. Everyone came. I had invited a cute single mother I knew named Barbara to come over to help entertain Norman. After the party she made it clear that she wanted to entertain me. Barbara was a very nice person, intelligent, and I was back in circulation.

In 1982 I met Julie Inwood. Her original family name was Parmakian and she had that legendary energy sometimes attributed to Armenians. Putting it bluntly, Julie had just dumped my colleague and great Diamond route climber Dave Rearick, and had selected me as her next climbing partner. Julie had known Rose slightly and was upfront right away with me in revealing that she had been diagnosed with an acute form of leukemia. They had given her two years and that was by then four years prior. I told her I was game, and game it was! Julie was nice looking with a winning smile, but her main attractiveness was in her athletic ability and energy. She loved to organize adventure trips and taught me scuba diving at Cayman Island. We took a two-week bike tour in southern France and then climbed the Matterhorn. We climbed many of the Colorado 14'ers together. We became very close.

Julie still had a high-school age daughter, Nancy, at home. I still had Patrick and Garth. Patrick was 16, but still, I did not like sleeping over at Julie's as often as she wished. And there was another problem. On her adventures, Julie would take great risks. Several times in exposed climbing situations I had to insist that we turn back. On a second diving trip to Cayman Island we went out alone to descend the beautiful 600-foot deep East Wall there. Julie "saw the yellow brick road" she had once described to me – when the beauty and the body's accumulating nitrogen tells you to just keep going down. I was also transfixed by the coral beauty of that East Wall, but then I realized I was steadily dropping deeper. I was barely able to quickly descend down to her and turn her around. It was very close, I estimate we were already down to 40 meters, which is more or less the "point of no return" for a single-tank scuba diver. Of course we had to also come up too fast as she was out of air. Eventually I told her I had to survive for my kids' sake, so I would have to back off from some of her adventures.

She easily found other male partners to replace me. But she always asked me first, for the next two years. Someone later asked me, "Were you soulmates?" I hesitated, then answered firmly, "Yes."

* * *

In March, 1985, on a fine bright day, five of us ski-mountaineered over Rogers Pass on the Continental Divide, and down to Winter Park on the other side. Julie planned such a crossing each winter, and some had turned into epics in blizzards and whiteouts. We had had a beautiful day this time, and all five of us settled into our single hotel room for the night. The plan was to ski back over the divide to the Eldora Ski resort the next day. We were Harlan Barton (a former boyfriend of Julie's), Julie, me, Julie's extremely strong skier friend Sonja, and Richard, whom Julie had been dating. She chose to sleep with me that night, and Richard had to take the folding cot. The next day we successfully skied back over the divide and were having drinks in the small Eldora resort bar. Julie sat on my lap and announced that she had scheduled a three-week summer trip to climb Huascaran in Peru and then raft down the Amazon. Would I go with her? I told her I didn't think I could make it. Within 15 minutes she was on Richard's lap. He agreed to go.

In April she went diving in the Caribbean with some partner, but on her return she immediately came up to a mathematics conference at Copper Mountain to join me for several days of skiing there. In May she called me for a bike ride up a local canyon road to the small mountain town of Jamestown. This is not a hard ride but I noticed that Julie was uncharacteristically lagging behind.

In June I received a frantic telephone call from her daughter Nancy. "Karl, Mom has died on the icefield of Huascaran." I told her to call her grandfather and aunt, both of whom live in Boulder. Then I called Harlan, who came over. We decided to call the hotel in the small town of Huaraz, which is the climbing base for Huascaran. The Swiss-born hotel owner answered. I established a little rapport by telling him about my living in Switzerland. Then I asked him what he knew.

It was no icefield accident. Julie and Richard had been out drinking and she had fallen in the shower, hitting her head. Later that night she lapsed into unconsciousness. Apparently Richard did not want to pay for the helicopter to Lima. After some delay, Julie's father paid and then flew down to Lima. A few days later, he made the decision to end her coma. The plug was pulled.

Julie had been my lover and friend. This was the second death of a soulmate. It also troubled me that I had not gone there with her – and Richard had. Perhaps due to the leukemia and medicines, Julie had an extremely low tolerance for alcohol. I knew this, and several times I had to pull her from bars where after only a drink or two she would get ornery and lose her balance. But I had already somewhat distanced myself psychologically because I knew the leukemia would get her eventually. Later we found out that she had come out of remission and had been quietly undergoing radiation treatments. That is why she lagged behind me on the May bike ride. On that ride, she had told me, "Remember Karl, someone here loved you." In her mind, it must have been prescient.

Three weeks later, I went ahead with the young climbing couple Ed Russell and Catherine Busch on our planned climb of the north face of Lone Eagle Peak west of Boulder. Julie was to have gone. Near the top of the face, after an exhausting climb of many hours, Ed led up a 5.9 vertical pitch. My fingers were tired, my eyes fatigued from looking for tiny holds all day, and my packstrap had broken and the pack was not well balanced on my back. I think I knew I was going to fall. I went for it anyway and came off and pendulumed about 20 feet to the left. I was not injured, but I had compromised the climbing party and was now hanging off-route. Ed called out, "Are you okay, Karl?" The wind was strong and I could almost not hear him. "Yes. Wait a minute. Just wait a minute!" I answered. I looked around and saw a crack system further to my left that I thought would go. It went, and I joined Ed. Catherine chose to follow my route, using a slow controlled pendulum over to it. I was 50 years old. That was my last rock climb.

Among my other lovers in those single-parent days was Ellen, who was a friend. She often fell in love with someone else, somewhere else – but then would come back to Boulder and we would go out. Also after Julie's death I dated a woman named Diane. Diane, a single parent with a son in my son's class, was one of the best lovers in my life. Diane was of Greek descent, had aquiline fine features and a lithe body, and thoroughly enjoyed sex with me. I wonder sometimes what makes the difference to a woman. I see myself as just your average lover. For a man there are not too many differences in the love acts, but I have concluded that for a woman, much is in her mind. I believe Diane fell in love with me, but not I with her. We were close but I represented more to her than she did to me. So it was not in the cards for her that I would settle down. I love her poems and the French language we always shared, and she still sends me postcards and letters from time to time.

By 1986 Patrick had joined the Navy and Garth was 14 years old. My friend Ellen hosted a jitterbug party and there I met Jill, a beguiling brunette. We danced, she squeezed my hand in one of those little signals, and I asked her to lunch. I was about to fall in love with an enigmatic heiress. Jill's family was an oil dynasty and had developed the Eldora ski resort.

In short order we went skiing a few times. Then on a snowy day we sought refuge in a private cabin Jill's family had retained at the resort. There she seduced me. I see now that I was "a piece of cake" for her. Afterwards she said to me, "That was the easy part, Karl. Now comes the hard part." But I had no idea what I was getting myself into. I thought I had found someone like Rose, and for the long term. We married soon after – Jill was insistent, and I was reluctant to disappoint her – and Garth and I moved into Jill's house. But something in my mind told me not to sell my house, so I rented it to a sabbatical visitor to the university.

Garth and I were warmly received by Jill's family. One of Jill's two daughters, Erica, took Garth under her wing. I thought it would be terrific. But I tend to not look for personal problems and am basically an optimist. In the beginning, I was absolutely captivated and in love with Jill. But I had seen only her good mood. I wonder if she rushed us into marriage intentionally, so I would not change my mind as I got to know her better, and saw her other side. I still think she is almost at the genius level. But when her dark side came, her cleverness became very cruel.

I soon learned that Jill had retained several of her previous lovers. Yet she was incredibly jealous and frantic whenever I even spoke to a female. Even one I had just met. "You want to have an affair with her!!!" she would scream accusingly. There were times she would say she was going to take an evening drive for an hour or two – alone in her sports car. When she came back I could smell cigarette smoke in her hair. Jill did not smoke. I knew she had not been alone. It almost broke my heart.

Jill had obliquely warned me. Early on she had told me: "Men run. That's what they do." She'd told me she had been married three times before, and that her last husband left after two weeks. I soon began to understand why, and then I was on the run too. Garth and I ran away to the upstairs apartment of my house, which was temporarily empty. I do not remember who did what, but in a few days, Garth and I returned to Jill. But I had decided that if we had to leave again, I would be prepared. I found a rental house that would be acceptable for six months, until we could move back into our house. A couple of months later when Jill, in a suddenly dark mood, told me, "Unless you and Garth are here to stay, I want you out this week." I was ready, and I got us out of there.

In January 1987, I was invited on short notice to come immediately to Paris to be a replacement Invited Speaker at a conference. Jill and I were still married, though living apart, and we were still sometime-lovers. In her good mood I could hardly resist this woman. She wanted to go to Paris with me. It was a hot time in the hotel room near the Arc d'Triumphe, and she really tried to maintain her good side. We had some really enjoyable times, and she was fascinating to be with. It was not all sexual. But then on the last day, I barely glanced at some pretty French woman getting off the Metro and Jill spun out of control and exploded at me. "You want to have an affair with that woman!" she shouted. She was very agitated. I realized at that moment that she really believed it.

The marriage was clearly over, but I just did not have the will to start divorce proceedings. Neither did Jill. One night she invited me out to dinner at a fine restaurant. I recall the moment, as a waiter filled her wine glass, when she gushed, "Karl, we can have it all! We can have it all!" I looked at my glass, which was only half full, and said, "Well, half-full is enough for me." She did not say anything but I could see the anger in her eyes.

Not long after, we divorced. There were a few power struggles at the end, but in November, 1987, I was again a free man.

* * *

My rushed decision to marry Rose was largely predicated on the needs of our five children. My divorce of Becky was certainly influenced by the needs as I perceived them of my little son Garth. When Garth and I "ran for our lives" from Jill's house (as I sometimes tell the story to close friends), indeed my decision had much to do with his welfare. Largely stuck down in her basement over the summer, I had equipped him with a computer terminal to play the computer games Rogue and Hack at the university, using my password. Near summer's end I noticed admiring stares from computer science students as I walked down the halls in the Engineering Building. Finally a group of them approached me and complimented me on having arrived at the very top of the competition. But it was all Garth, at age 14. The revelation struck me: something was wrong with that picture. So the fate of all three of my marriages has turned on the needs of children. I can be proud of that. I have never been a naturally selfish person.

My first wife Becky remained in Boulder and was a responsible and caring mother to our children. In her late teens, my daughter Amy moved to her mother's house. Many years later Becky remarried and her second husband and I are on good terms. Sometimes we are all together at some activity of the children. To my knowledge Jill has not remarried. We currently all three live within a mile of each other in South Boulder.

In 1990 I met Kathy, an agile athletic woman with whom I have had great fun during the last twenty years – hiking, climbing, skiing, and sometimes as a lover. The main thing has been the friendship. Kathy is an independent soul, has never married and greatly enjoys outdoor activities. Although we are in and out of touch, Kathy remains a long-term friend. Karin, another skiing and hiking companion I have known for many years, also remains a valued and trusted friend.

I must also fondly acknowledge my great friend Jacqueline in Brussels. We met at a Fulbright Scholars rendezvous in Mons, Belgium, in 1979. Her boyfriend Alex had gone away for 3 years to law school in England. Jacqueline invited me to keep her company at her flat. I accepted. Recently she reminded me of how it went that first evening as we prepared dinner together. She placed Alex's picture, and I placed Rose's picture, on the mantle above the dining room table. It was my first experience at agreeing to be close but platonic friends with an attractive woman. I enjoyed Jacqueline's hospitality, friendship, and nurturing in Brussels for 25 years. Her place became my home away from home. If her job as an EU translator took her away for some weeks, I knew where the key was. Eventually Jacqueline married, but we will remain trusted friends for life. She always wants me to visit, and I did a year ago at their summer home in France.

* * *

Why do you love someone? Why does someone love you? Why are others friends? I really cannot answer that. But I would conjecture that much lies in a hidden chemistry which rather instantly determines the outcome. This chemistry cannot be turned on and off at will. It is a part of your total being. It involves your whole identity.

Have I been too susceptible to beautiful women? I think not. They have in most cases, either directly or in more subtle ways, invited me into their lives. However, their expectations were sometimes not the same as mine. It seems that all of my life, the order of priorities in my personal life has been friends, lovers, wives – and not the other way around. Lovers come and go. And marriages are for the kids. The key has always been the friendship. That will carry you through.

Notes

Rose's oldest son Christopher went to college and earned a Master's degree in Environmental Engineering. A champion runner and currently active in cyclocross, he lives in Boulder and is Uncle Chris to my grandchildren. Rose's daughter Kelley had all the traits of a runaway, and even our transferring custody to her father did not prevent that. Sadly, Kelley died at age 44 in 2008 of the identical cancer that killed her mother at age 42. Patrick lives with Chuck, his partner of fifteen years, in

Oregon. My daughter Amy was heading toward a Ph.D. in History but stopped at the Master's level, and then earned a law degree from Berkeley. She is a successful lawyer in the San Francisco Bay area where she lives with her young family. My son Garth is Parts Manager at a local auto repair establishment, and I have been able to greatly enjoy his young family's proximity.

10. Close Calls

...in particular, the lightning strike...

Most of us tend to take it for granted when we arise in the morning that we will live through the day. Even knowing that there is no certainty of that, why spend precious time considering what possibilities will confirm our mortality? But few of us reach the "real event" without some close calls to prepare us, moments where we are given a preview of the fine line that lies between life and death...and a chance to be grateful for the ability to greet another day.

In my climbing days, there were many close calls. Maybe that was part of the appeal of climbing – though mostly they were manageable events. I already detailed in Chapter 2, "The Boy in Boulder" the perilous night in our tents at -60 degree on Mt. Lincoln. Survival was in the works there because we were a hardened, tough team of teenagers. Another close call was my final pendulum on the face of Lone Eagle Peak in 1985, recounted in Chapter 9, "Wives, Lovers, Friends". There were other tense times, where I had to wonder, "Will we make it out of here?"

Here is one such incident I've never told anyone about. In 1953 we had decided to do the Window route on the East Face of Long's Peak in Rocky Mountain National Park. Ours would be the third ascent of that route and in fact after we went through the Window, a small hole in a fin of rock jutting perpendicularly out from the East Face, we would put in a new first ascent on the other side. The previous summer we had already climbed Stettner's Ledges, a hard route below the Broadway ledge that runs across the entire face. However, this was our first venture onto the steeper Diamond portion of the East Face above the Broadway ledge. I was leading the first rope as Skip and I moved off Broadway to angle up to the bottom of the Window slab. I had no idea if I was on-route – indeed, there was not yet any established route. Although the incline was nearly vertical, I had not bothered with placing any protection yet because we were just on the approach to the crux of the climb. And in those days, we were always fast, agile climbers.

I can still feel today that instant when I thought I was coming off the face. It was on some small ledges, not really ledges, but the kind of little 60 degree few-inch variations in an 80 degree face that one can put hands or feet on to inch up. One might call it near-vertical crawling. You don't have any real holds. You just have these rugosities to work with. So you may, for example, put the heel of one hand on this little portion of less gradient, while balancing with your feet and other hand on similar little lesser gradients. It is like the friction climbing we always did on the rocks around Boulder. But there the gradients are lower and by pressing down and moving delicately, one is staying on the rock face primarily through the friction of one's climbing shoes. Here, I needed the foot friction plus that in the heels of both of my hands.

It was not enough. Had I come off, both Skip Green and I would have flown off the face to bounce once on Broadway and then continue down the lower face to our deaths on the rocks and snowfield at the bottom of the East Face. I don't know how I stayed on. I had the feeling of coming off. It was like my breath had been taken away. But somehow adrenalin had poured into the heels of my hands to push me a little tighter into the rock, to stay on. So I did not come off. I never told anyone about it that day, or later.

The next brush with death that I recall came from our terrible habit of racing cars in high school. We called it *ditch-em*. Some of the gang would be in one car and some in another. One of our favorite locations was the Flagstaff Mountain road above Boulder. It was gravel then, and I can still feel the sliding around the corners at high speed, hoping to lose the car behind. But one dark rainy night we were racing east on Baseline. Lynn Ridsdale was driving his old heap, I cannot remember the car make, and I was in the passenger seat. We were trying to ditch Jim Vickery et al. in the car behind. We turned left onto Cherryvale Road and headed straight north at high speed. Suddenly I felt a huge impact, then a car full of dust flying forward at high speed, bouncing off trees. We came to a stop in complete darkness. Lynn was able to get his driver's door open and fall out on that side. Somehow I extricated myself, made my way over to his side, and fell out on the grass beside him. A few minutes later Jim Vickery and someone else were there beside us. Both Lynn and I were in shock but not seriously hurt.

What had happened was very simple. We were not familiar with the roads out there. In those days it was farm country. Cherryvale Road dead-ends onto Arapahoe Avenue, and at 80 miles an hour we flew across Arapahoe, hit a tree on the other side with the right front of Lynn's car, ricocheted almost perpendicularly to the left to hit another tree, then continued straight north through two farmers' fences until we came to a halt. Miraculously, the car had not flipped. At the impact with the first tree, my knee had crushed the glove compartment, my foot had buried itself into those little car heaters one had under the dash in those days, my hand had bent the dash panel inward, and my head had cracked the windshield. The car was totaled. Of course Lynn was punished by his dad, and that was the last game of *ditch-em*. If our trajectory across Arapahoe Road at 80 miles per hour had been a couple of inches more to the right, I probably would be dead.

Generations succeed generations. Or they do not. I must insert here an analogous near-end of my son Garth. I don't know all the details. But in high school he bought an old 1967 Chevy Camaro with 210,000 miles and completely rebuilt it. He learned a lot about automobiles in that effort and I supported it. He basically rebuilt the whole car. Of course the Camaro is one of those classic muscle-cars that teenagers believe are great symbols of prowess. But his beautiful Camaro nearly killed him. How do I know? One morning I went out to our driveway, and there was his Camaro with the passenger side completely caved in. Garth had wrapped it around a light-pole at very high speed. Luckily he had been driving alone. Somehow he managed to drive it home. I found him completely passed out sleeping the sleep of emotional exhaustion in his room. Partly out of memories of my own teenage years, and partly because I love him, I never pressed for details.

Another automobile-related close call nearly killed me in my youth. I had graduated from high school and was a Trail Crew foreman at Rocky Mountain National Park for the summer before starting my studies at the University of Colorado. I had an old used car with old bare tires which were always blowing out. In those days, you jacked the car up and patched the tire yourself, or put on the spare if it was a full blowout. My left rear tire had blown as I was driving down from Estes Park to Boulder. It was on the portion of highway just a mile or two above the town of Lyons. It was late at night and I had only a very small flashlight. So there I was, on a small shoulder of the road with my body surely out onto the pavement, trying to jack the car up and replace the tire with the spare. I can still feel the rush of air on my back as a large truck hurtled out of the dark around the blind corner and past me. I would estimate that he missed me by an inch or two. They say that football is a game of inches. But so is life.

I was pretty busy raising the two families and staying out of trouble in the middle portion of my life, and no near-death events come to mind. But later in my life, there were two episodes. The first is my escape, by seconds in time rather than by inches in space, from death by lightning.

Norm Nesbit and I had flown out to San Francisco to join our old climbing buddy Bill Bueler for our annual "old guys" outing in 1997. After some scrambles and a climb in Yosemite, Bill headed back west to Monterey and Norm and I continued east to Owens Valley. Our idea was to do a very easy California 14,000er. I had never done one in California. Norm had only done Mt. Whitney about 40 years earlier. We picked White Mountain, a walk-up, even a drive-up if you have a good Jeep. We drove to the road gate at 11,000 feet and camped for the night. We started after breakfast at about 7 a.m. Here is my account of the incident, which I wrote shortly afterward.

* * *

Three Frames

We were nearing the summit of White Mountain, the 14,246-foot high point of the White Mountain and Inyo Mountain Ranges, which separate the Owens Valley from Death Valley near the California-Nevada border. There were just two of us and we had risen early to the strange sight of clouds around and below us to the east. Strange for a desert mountain range on which ancient bristlecone forests kept their timeless vigil praying for a little moisture. Strange clouds, for our climbing friends had told us that such do not occur so early in the day in California, as they sometimes do in Colorado. So we had started out at once, and at a fast pace we had gone in three hours from camp almost now to the summit. It was only 10:00 in the morning on July 28, 1997.

White Mountain is a walk-up, with a trail that's an old jeep road rising all the way to the summit. We had done a couple of harder climbs in Yosemite, and just wanted this one as a "14000er" and for nice views of the Sierras to the west. We both had climbed all the 14,000-foot peaks in Colorado and had a lot of experience with and fear of lightning. So we paused one last time to talk about our situation. We had heard absolutely no thunder at all, anywhere. We had seen no lightning, anywhere. So we continued up, fast, cutting straight up the small boulder field directly toward the summit. There were even a few sunlit patches on the slope below us to the northeast, and some blue sky to the west of the summit. The danger lay behind us to the southeast, the same strange dark cloud bands building out of Death Valley, due to the rather unusual gulf monsoon weather which was penetrating the southwestern states.

As we moved quickly up to the summit, it suddenly began to hail on us. I touched the top at 10:15 am and after about a minute started down. Norm was just topping out and I told him there was a lot of metal on the small research hut on top and that I had heard some "bees": that hum of static electrical discharge well known to all mountaineers. He veered to his left and immediately started down parallel to me, about fifty feet to the south. It was 10:17 a.m.

Frame 1. I am surrounded by warmth and light. As if in a dream, I am hurtling downward. Incredibly, by athletic reflex I am staying upright as my boots fly down over the small boulders. There is no sense of fear or pain but instead a sense of wonder and almost well-being. There is also a clear feeling of absolutely no control over my body.

Frame 2. This frame is blank as I momentarily lose consciousness.

Frame 3. I regain consciousness and I am lying in a pile of rocks and my right leg is paralyzed and numb with pain and I cannot move it as I try to get up. I do manage to get up but then fall down.

I look to my right and see that Norm is continuing to move rapidly down the boulder slope. It is hailing hard and I know more lightning strikes will come. I try again and find I am able to get up and swing my right leg. Better, it has some feeling again, although my thigh feels strangely hard. Still, I can move slowly downhill. I stay low and make my way down the boulders to Norm, who is crouching in a small saddle at 13,500 feet. There are numerous lightning strikes around us. We hunker down in a small low place under our ponchos and wait. After thirty minutes the first series of blasts has moved north of us and we emerge to about one inch of small hail of graupel size covering the mountain as far as we can see. It is snowing. The temperature has dropped dramatically.

There are still ominous sounds and blackness over the ridge to our left, along which the trail descends, so we avoid it and side-hill south. As we head over the last little divide down to the University of California research station five miles below the summit, we are caught in a series of lightning strikes coming off the ridge just to our left. But because there is no cover and we see no low places, we just crouch low and continue on. At the research station the guardian tells us it is the worst electrical storm he has seen in his six years there. Even as we continue on down to camp, we are in a blizzard again.

On the long drive to San Francisco and flight back to Colorado, my leg is stiff and aching but tolerable. Then the next day the hematoma, all the ruptured little blood lines, begin to show. They cover my leg and especially the back of my thigh. No burns, but clearly I took some amps.

One can do everything right, at least within limits deemed reasonable by experience, and still get hit. It can happen on a mountain, on a bike, in an automobile, in a home. The timing of unexpected fortune and misfortune is somehow beyond us.

* * *

As I recalled the lightning strike some time later, certain things became clearer. First, I recall hearing a loud "boom" just before Frame 1. That would mean that I got the second strike. Lightning often has several strikes in its discharges between earth and sky. If that was the case, not only was my life saved by staying only a very short time on the summit, but my warning to Norm may have saved his life too. The first bolt that hit the summit must have been massive due to the nature of this unusual storm's long buildup. Second, I was hit mostly by the ground stroke. By that I mean that a lightning bolt arcs from both the sky and the ground. That ground stroke ruptured my leg. Clearly the electricity also flowed through my brain. That was the beginning of Frame 1. Although I did not write it down then, I did see that "lovely white tunnel" that seems to be common to such near death encounters. I can tell you that it really does feel like Heaven.

The latter part of Frame 1 and Frame 2 represent my being thrown downward over the rocks by the blast. I could guess it blew me about ten feet down the mountain, but that is just a guess. My feeling of great athletic prowess is pure illusion. Frame 3 can be thought of as the beginning of Hell. Maybe it would have been better to just stay permanently in that Frame 1, Heaven? Surely I would choose that over other ways to die.

There was a lot of pain although of the dull aching variety. Also our perilous descent for three hours, going slowly because of me, in the middle of a fierce lightning storm on a high exposed ridge, was like being in a combat zone with mortar strikes all around you. There is a memory of how our stoic mountaineer natures reacted to the situation. When I staggered down to Norm at the little depression at 13,500 feet, he was already crouched in the recommended posture, with his parka over his head. He peered up as I said, "You know, I think I got hit." There was that mountaineer's pause, then he said, "Can you walk?" Another mountaineer's pause, then I said, "I think so." End of conversation for awhile, while the storm raged on top of us.

We had a long drive back to San Francisco and even had to overnight once to get there. In spite of the dull but constant throbbing pain, I did not want to go to a doctor until I was back in Boulder. Besides, there were no burn marks. Then the leg turned black. When I finally called my HMO and said I had to come in because I had been hit by lightning, my normally reticent doctor got excited, said come in immediately, and when I arrived, I found all three doctors there eagerly waiting to look at me.

There is a little bit of academic curiosity in all of us. The skin surface blackness was removed in about three weeks of electrolysis treatments. When they started that treatment the first day, they wrapped my leg and then slowly turned on the current. The dial moved up to about 50 and they stopped and asked me if I felt anything yet? No. They slowly moved the dial up to 100 and still I had not felt anything. They consulted and decided to let me control the dial. As I recall I took it to the highest value and still felt nothing. My skin was dead. It took about four months for the ruptured interior leg muscles to recover.

* * *

My last close call to be recounted here came as a result of my own stubbornness. It was in October 2007, and there was a severe flu going around. I was teaching two courses. Approaching was a Thanksgiving break week. On Wednesday of the week prior, I was to give a midterm exam in one of my classes. On Sunday, I did not feel well. In fact, I felt terrible, but I attributed it to stomach flu. On Monday I was scheduled to serve from 4 to 5 p.m. on a Master's Exam presentation, and I did not want to miss that, for the sake of the student for whom it was an important event. So I showed up Monday, taught my classes, and felt worse and worse. I told the student I would be at his exam, but that I needed to rest in my office, please wake me up if need be. It seemed natural at the time to just lie down on the floor in my office, although certainly in retrospect that seems a strange decision. The student dutifully knocked on my door and I went to his exam. Midway in it I had to excuse myself and go the bathroom to throw up. I almost never throw up. After the exam finished, I managed to drive home, feeling really exhausted.

It is a danger to live alone. Had anyone been there with me, they would have insisted I see a doctor. But I don't like to go to the doctor. I will always remember my wife Rose's advice to me as she was being treated for cancer. She said, "Karl, stay out of the medical system if you can." She was putting some humor into her own situation. Had she been around that Monday night she would have rushed me to the hospital. So I just toughed it out but could not sleep due to the pain in my abdomen. It moved around and I wondered if I had some strange kind of blockage, like constipation. Finally, at about 2 a.m., suddenly I felt a little bit better. I got an hour or two of sleep and made it through Tuesday. Tuesday night I somehow slept a few hours and went to my first class Wednesday to give the mid-term exam. I felt exhausted and had to ask two students to pass out the question sheet and the exam paper. This had been a good bunch of students and no one complained as I just sat at the front desk in near collapse as they worked and each handed in their exam paper. Then I managed to walk to the Engineering building for my graduate course. It was a cold day and evidently I just walked in and plopped down in the front chair with my coat and hat still on. A retired physicist, Steve Nerney, who was auditing the course, a very positive fellow who I had really enjoyed having in the course, looked at me and quipped, "Uh-oh. Doesn't look so good – you're not even taking your hat off!" He knew something was wrong. Another student said, "Professor Gustafson – you look terrible – can I drive you to the hospital?" I replied, "No thanks, I think I can make it myself. But no class today."

I did then manage to drive to the local Kaiser clinic. That was about 1:30 p.m. Some blood tests were done and an ambulance called. In the meantime they put me on an I.V. table and hooked me up. The ambulance did not arrive until about 4 p.m. When I learned they were sending me to a hospital, I was incredulous. I managed to talk a nurse into going out to my truck and retrieving my backpack, which had those exam papers in it. I was operated on at about 10 p.m. By now it was evident that I had been walking around with a burst appendix for about 40 hours. The sensation of "feeling better" 40 hours earlier was the appendix bursting. The operation went well, but it took a four-night stay in the hospital to drain all the poison from my body.

This was not a close call in the sense of inches or seconds. Nonetheless, statistics show that for stubborn old folks of my age, about 25 percent die when their appendices rupture and they do not seek treatment. Those are the same death-odds resulting from a lightning strike. Although the odds that your appendix will rupture depend partially on genetics, exactly when it will rupture is a matter of chance. Whether you will die of a lightning strike depends partially on the risks to which you will expose yourself, but also the chance as to where the bolt chooses to strike.

Even the quality of my treatment once I was finally delivered to the hospital may have been influenced by chance. I had been told a year before that my former Ph.D. student Shelly Goggin had married one of the surgeons at this hospital. A Doctor Varner walked in at 10 p.m. that night to chat with me just before cutting me open. Although I had been on the happy I.V. stuff for 7 hour already, I told the anesthesiologist just before she put me away, "Hey, do you know Dr. Gerding or his wife Shelly?" Dr. Varner immediately perked up and said, "Sure, I know Shelly." I explained that she had been my Ph.D. student and we had worked together for five years. I find it very credible that I received an extra-fine operation from Dr. Varner, no huge incision but only three small entries and a drain hole, partially because of my past trajectories-link with Shelly.

I am convinced that when you recognize how our own individual life's trajectory is so interwoven with many others, fate is playing a game of chance with us, even at every instant. In extreme defense of that assertion, just realize that as you read this, swirling about you in the room are the trajectories of literally millions of viruses, some potentially lethal. Our instantaneous being is always in a perilous dance between life and death. One need not be freezing on a mountain or losing to gravity on a rock face or racing cars or changing flats in the dark or being hit by lightning to be close to death. Those are just the easily recognized close calls. They are the close calls where we even may think that our own quick choices or strength or instinctive reactions made the difference between life and death. There are many other close calls, for example a momentary lapse by a driver coming at you at high speed, corrected only by her last-minute swerve, that never enter your awareness.

Life and death are poised in a delicate balance for all of us. We live in a sea of such uncertainty. I am not too concerned about my own mortality. I have never been a self-centered person so I don't place any special importance on my own life as compared to the lives of others. But when an event of good luck happens to come to you, be sure to recognize it, enjoy it, and do your very best with it. My particular luck at surviving the lightning strike changed my life. That moment at 10:17 a.m., on July 28, 1997, is etched into my mind. Whenever it rises to my consciousness, I smile and feel very alive. Each day since has been richer and more fascinating.

Notes

Skip Greene was purposely flirting with death in his solo East Face winter climb (*see* Chapter 2, "The Boy in Boulder"), but he survived. He never knew how close he was to death when I nearly came off on our Window climb. Ultimately, he died of a very fast-acting leukemia on Jan. 10, 2010, during the writing of this book.

11. Mathematics

...and some science...

Mathematics is akin to a foreign language to those who do not know it. It is not really possible to become fluent in a language unless you are long-immersed within it. When asked what I do or teach and I answer, "Mathematics," the reaction is often a disinterested, "Oh." That is a shame: mathematics can be beautiful. And of course it is useful.

Some have called mathematics the handmaiden of science. I have seen too many instances of the superiority of mathematics to science to accept the tacit subordination of mathematics implicit within that statement. In my work, mathematics and science have walked hand in hand when pursued in combination. Each has needed the other, in an essentially equal balance. So in my view, it is alright to say science needs mathematics, but one must keep equally in mind that mathematics has a need for science.

On the other hand, mathematics can have sufficient value to abide entirely within itself. Thus I am not at all at odds with those who view themselves as pure mathematicians and who need no further justification or inspiration other than that which comes from the body of mathematics itself.

To elaborate this point, let us put all of mathematics into a box: A box has the three dimensions of breadth (width), depth, and height. Now let us put any individual mathematician's little sub-box into the big box. What is the extent of his/her little box within the big box?

In this picture, a typical pure mathematician's little box will be an extremely thin slice of tiny breadth, some small height as he/she accumulates research and scholarship over the years, and relatively large depth as the individual digs deeper and deeper into his specialty. A typical applied mathematician will need more breadth to his little box for application purposes, more height due to a larger accumulated knowledge base over the years, but less depth because of a need to keep moving on to new problems. A mathematical physicist will need even more breadth in the sense that applied mathematics usually just focuses on one aspect of the physics, more height because so many mathematical and physical tools will be needed to properly pose the problem, and more depth than an applied mathematician but perhaps less depth than a pure mathematician.

Thus in my assessment, I order the three classes by extent and find that generally

$$\text{pure} \leq \text{applied} \leq \text{math-physics}.$$

Of course there will be heard howls of protests! Let them howl. I have successfully done all three kinds of mathematics.

A mathematical logician (very narrow field) quite a number of years ago complained to one of my colleagues about his publishing a great paper that year and getting a pay raise that was lower than mine. My colleague looked and saw that that year I had published ten papers. To the "my-one-is-so-great" logician my colleague simply replied, "Karl has three great papers in there, maybe you should publish ten papers next year!" Such arguments can go on and on...but the logician retired, and the three of us now get along very well over at the campus gym. The fight for prestige vanished upon his retirement.

I prefer to defend all three proclivities. Mathematical research no matter of what type is quite demanding. There is no room for fudging or ambiguity or imprecision. And the world-wide competition that the young assistant professor now faces is intimidating. I was very lucky to have entered mathematical research in the expansive period of the 1960s when the field was blossoming. A useful analogy is that of rock climbing. I could do a 5.8 rock route back then and get a first ascent. Nowadays world-class climbers come to the Boulder area and fight over 5.13^+ routes. The beautiful fun routes we did are trivial. The low-hanging fruit is all picked. The competition now is between technical specialists.

Because of this, I am saddened to see our younger colleagues often possessing almost no breadth at all. They think they are deep but I see them as highly trained technicians. Another handicap is their lack of training in physics and engineering. Again however I think they need not apologize. Mathematics has narrowed its scope, of necessity. Engineers, physicists, economists have become their own applied mathematicians. One cannot know all fields, or even all of one field in mathematics anymore.

* * *

For me, mathematics was always easy in school, but I had no particular passion for it. Moreover, mathematics was a very tiny profession in 1953. At that time as a senior in high school I was advised to enroll in engineering at a university, and preferably in chemical engineering, as there one would find the highest salaries upon graduation. That I did. But I quickly found chemical engineering as a technology quite boring. Also I did not enjoy the endless hours in chemistry labs. I did not like learning things by rote rather than by reason. Finally, I did not like the idea of doing science in order to join a corporate structure or to form my own company to make a lot of money. Engineers, more than physicists and mathematicians, often have minds not unlike business entrepreneurs.

My transfer from chemical engineering to the engineering physics major occurred in a personal way. I was complaining to Dick Becker, with whom I had done the first ascent of the north face of Mt. Meeker after my sophomore year at CU. Becker was two years ahead of me and was completing an engineering physics major. "Karl, why don't you just go into engineering physics? It is the most difficult major on campus." So I did. Indeed, it was a tough major. I remember my E & M (Electricity and Magnetism) class in which 25 started and 5 finished. I took and survived all the required courses except the Electronics and Radio course, which I missed when I was switched over to the Applied Mathematics 1958 graduating class in my last year. Therein is a small irony, since within a year I would find myself in the Radio Division in one of the great electronics research laboratories of the world, NRL in Washington, D.C. No matter; I was never much good as a lab rat. I seem better suited to working with ideas and with pencil and paper.

The engineering physics majors were a special breed. They were few, and gloried in the prestige of being such, much like those in the Marines in the Navy. They joked that if you could not make it in engineering physics, you dropped down to just a physics major, and if you could not make it there, you dropped further down to become just a mathematics major!

As I reflect upon it now, I cannot entirely shake off that view.

The upshot is that I have a much higher appreciation of physics, applied mathematics, and even engineering mathematics, than do most pure mathematicians. I am not at all afraid to do mathematics and at the same time its applications such as physics. Most pure mathematicians do not have enough experience to have gained enough intuition and confidence to be as versatile.

On the other side, pure mathematics possesses a kind of absolute truth that one does not find in the other branches of science. A mathematical theory or theorem, once carefully proven and independently verified by others, is correct in and of itself. Its truth is a matter of pure thought and depends only upon the axioms that preceded it and upon which the steps of the proofs of the theorems were based. The theorem may have no applications, or many.

Those of us who have done a lot of interdisciplinary mathematics therefore have roughly equal appreciation of both pure mathematics and of its fields of application. I must emphasize that I never take lightly the idea of applying some pure mathematics that I have created myself or even applying some standard pure mathematics to some new field of application. I tell students in my mathematics courses when often I skip the so-called sections on applications added at the ends of chapters: "One does not do an application in a day. One does it in several years. Otherwise you do not know it, and are cheating it." In those instances when I have successfully worked interdisciplinarily, I have taken the view that I must understand that other area, be it physics, aeronautical engineering, neuroscience, optics, or finance, as well as do those scientists who are experts within those fields. It is a tall order. Few can do it. What helps is a general intuition for how to do so formed from having done so before.

I will always have a strong, natural, intuition for mathematical computation and some aspects of computer science, just because of my total immersion in it during my NRL work for the military early in my career. Then my Ph.D. training in partial differential equations at the University of Maryland gave me a solid handle on what is commonly called classical analysis. In my last year there, when I had already essentially obtained the needed classical results for my dissertation, I learned the more theoretical treatment of partial differential equations, called functional analysis, a field of mathematics in which I have worked ever since. My graduate mathematical training also included considerable topology but I am not much drawn to that subject, or to pure geometry, or to pure algebra. Thus if you would force each mathematician into just one of the three categories of Analysis, Algebra, Geometry, I am in the first.

* * *

Where have my mathematical interests taken me? What are some of the results? Can we describe them here in non-technical terms? I think so, if you permit me just a little latitude in exposition.

To begin, my Ph.D. research in partial differential equations required that I make substantial use of the mathematical theory of inequalities. That was true also for John Nash when he obtained the famous De Giorgi-Nash estimates for partial differential equations. The term estimates in differential equations usually is synonymous with the word inequalities. Nash has said that had not De Giorgi also obtained those inequalities, he, John Nash, would probably have been awarded the Fields Medal in mathematical research.

However, I have not really published many papers in theoretical partial differential equations. I am known in the field largely because of my book on PDEs originally published in 1980, later translated into Japanese, and now in its third Edition. And I have almost continuously taught PDE courses at CU, albeit only at the lower levels after the departments split in 1990.

My post-Ph.D. interests turned to mathematical physics, and a part of me has stayed there ever since. My first paper, and then many soon after, established rigorous theorems and facts of a functional-analytic mathematical nature needed by theoretical physicists working in quantum theory. One found in those days that those of us interested in quantum physics were about equally split between the physics and mathematics departments. Nowadays you find very few in departments of mathematics.

In the late 1970s, the job market for Ph.D.s in theoretical physics and mathematical physics dried up precipitously. Meanwhile, the roles and needs of computational physics and computational engineering suddenly accelerated. Because I already had some experience and comfort in the general computing environment, several engineering departments approached me to teach courses in computational differential equations to which they could send their graduate students. I initiated and taught such courses throughout the 1980s until the engineering departments could hire their own computational faculty. As a result, I produced several Ph.D. students within several engineering departments. With those students I wrote numerous original research papers. Of necessity, I became a state-of-the-art aeronautical engineer. With an outstanding Ph.D. student, Robert Leben, we became the first to successfully model and numerically compute solely by partial differential equations coupled to fundamental principles of aerodynamics, the dynamical motions of dragonfly flight. We also developed innovative mathematical algorithms which could capture the true multi-vortex lift development over an airfoil. We tracked fluid vortices into a square corner to such precision that we could resolve more than twenty of them. Those were sharp results in computational fluid dynamics in that era.

In the late 1980s I was approached to enter into a seven-department bid to bring to CU a $22 million NSF engineering research center in optical and neural computing. I was the only mathematician involved. We succeeded, and for five years I plunged into neuroscience and mathematical algorithms which could be implemented opto-electronically. This was fascinating work, more in the genre of engineering and in which one needed to be also competent in physics and computing. I greatly enjoyed it even though the goals were operating systems rather than abstract theorems. In such research a team effort is requisite, and to my tastes, rewarding. However the bottom line was that with rare exceptions, we could not invent new non-digital computing paradigms which for general purpose computing were superior to the existing digital computer industry. As a result of that experience, I am prone to say, "It is hard to beat digital!" Indeed, having been a digital computing pioneer in the early 1960s, I am proud and awed by our modern digital world only fifty years later.

When the Mathematics Department was rendered asunder in 1990, I went back to pure mathematics. Having seen within a couple of years that I had been closed out of any grant funding in applied mathematics, and having been too long absent from the funding mafias in pure mathematics, I decided to go back on an unfunded basis to an early creation of mine in the years 1967–1972, a theory of matrix and operator antieigenvalues and antieigenvectors. I have now published more than 70 papers in this mathematical subject which I created, and I am often invited internationally to speak on that theory and its applications.

To elaborate somewhat, it should be recalled that much of mathematical analysis and engineering and physics take a given situation and analyze it by breaking it into its fundamental modes, whether they be wave components, vibration frequencies, or atomic harmonics. The resulting fundamental physical or mathematical numbers are called the *eigenvalues* of the problem, and the physical or mathematical modes which generate them are called the corresponding *eigenvectors* or *eigenfunctions*. The German word *eigen* connotes self-mode or inherent-mode, and the *eigenmodes* are those natural modes which self-reproduce under the actions of the physical or mathematical operator. Here's a layman's example: when you drive on a rough gravel road and suddenly your car starts shaking and vibrating in resonance with a certain pattern of ripples in the road surface, the driver says, "Whoops, we are 'wash-boarding'!" The engineer says, "We hit a critical vibration frequency." The mathematician says, "We are at an eigenvalue."

My theory of antieigenvalues was induced by a problem in the mathematical theory of stochastic processes in 1967. After some analysis I reached a point where I needed to know not the critical eigenvectors of a mathematical operator, that is, not the operator's self-reproducing modes, but instead I needed the modes most turned by the operator. These I viewed as the least self-reproducing, and accordingly I called those modes the *antieigenvectors* of the operator. To the cosines of the maximal angles turned I gave the name *antieigenvalues*. I have applied my antieigenvalue-antieigenvector theory to many domains, including stochastic processes, numerical analysis, wavelets, signal processing, Bell's inequalities in physics, and recently to a number of problems in matrix statistics, quantum geometry, and finance. It has been great fun.

* * *

Looking at my 20 Ph.D. students provides an overall gauge of where I have spent my academic research mentoring efforts: Four students have done original research in functional analysis; three in PDE; seven in numerical analysis; four in mathematical physics; and two in mathematical finance. Of these, seven took Ph.D. degrees in Pure Mathematics, seven in Applied Mathematics, two in Physics, and four in Engineering: one each in the Aeronautical, Chemical, Civil, and Electrical Engineering Departments.

* * *

While visiting the Asea Brown Boveri (ABB) research laboratory in Baden, Switzerland, in the early 1990s for some collaborative research on neural-net simulation of stock market dynamics, one of the management staff by chance sat down next to me at the cafeteria. After introductions, we embarked on some really stimulating scientific discussions related to my research. Then he returned to his management mode and complained, "You know, Professor Gustafson, I have to give back 2 percent of all my contracts so that they can bring in finite element and other numerical experts here." He was astonished and chagrined when I responded that it was those budgets that paid my visiting expenses! Generally speaking, mathematicians are put at a disadvantage to other scientists when it comes to budgets.

* * *

It seems to me that there is a strange paradox that underlies not only mathematics but also physics and even engineering. For mathematics I may state it as follows. *The Generalization-Specialization Paradox*: Each time you generalize a theorem, you lose generality. What do I mean by this?

In pure mathematics one goal is to take a known theorem and make it a more general statement. Often this amounts to showing that one of the previous assumptions may be weakened or even thrown out. So in principle you have obtained something which may have wider validity than before.

But I have noticed that as mathematical theorems and theories become more general, they become less interesting to practitioners, and are used less. Thus generality of use has been decreased. It is actually the so-called engineering details which are the essence of any use of the theorem. Moreover often it is in those details that further true mathematical beauty lies.

A well-known historical example I may call on to illustrate this paradox is the rise and fall of the Bourbaki School in France. A brilliant group of French mathematicians decided to create a new edifice for mathematics which was strongly hierarchical, internally consistent, efficient, and clean. To do so they decided to keep it free of any tainting of applied mathematics, and to not be influenced in any way by applications or other parts of science. Certainly their achievements in so doing were remarkable. But the increasing abstractness incurred a collateral decreasing practical usability of their structures. A famous leader of the effort, Alexander Grothendieck, suddenly dropped out and disappeared into a hermit life in the Pyrenees, saying in effect, "We went too far." Grothendieck then wrote a great diatribe of more than 1,000 pages, which I went to considerable effort to obtain for our library here. It is a fascinating (albeit at times depressing) read.

I was trying to explain my Generalization-Specialization paradox to a delightful younger colleague of Russian-Jewish ancestry who sometimes passes by my house on his evening walk. Definitely of the pure mathematician persuasion, he charmingly did not agree, and we went on to discuss other topics as dusk settled upon us. To illustrate one of them, Sasha volunteered the following anecdote: The peasant goes to the rabbi and says his chickens are dying. The rabbi instructs him to draw a large circle on the ground and put the chicken-feed inside it. A few days later the peasant tells the rabbi that his chickens are still dying. The rabbi tells him to draw a large square in the dirt and put the feed inside it. After a few days, in desperation the peasant informs the rabbi that he has only one chicken left. The rabbi tells him to draw a perfect five-point star and put the feed inside it. The next day, the peasant returns, crying, and tells the rabbi that his last chicken has died. "Oh, that's too bad!" the rabbi exclaims. "I had even more beautiful geometries to try!"

I looked directly as Sasha and said: "Don't you see, Sasha? Your story illustrates perfectly the dangers of too much abstraction!" As he gazed back at me, it plainly dawned on him that, indeed, it did.

Mathematicians of course still line up on both sides of the issue of depth versus breadth. Most pure mathematicians value depth more. Most of them possess little breadth. Therefore you cannot convince them otherwise. In fact, they believe that because they are abstract, they are broad. But in the real world, you need breadth of competence, rather than abstraction; and you need breadth of experience in related fields, rather than just theoretical breadth in your specialty.

My generation of mathematicians, physicists and engineers was privileged to be exposed to more breadth of subject matter and work experience than is the present generation. First of all, this was required, or at least strongly encouraged. Second, it was more possible. As scientific fields advance, they bifurcate into specialty areas of expertise which demand all of one's time. Look at Medicine. One makes far more money by specializing. General practitioners and house calls are a thing of the past.

So the Generalization-Specialization Paradox is not found solely in mathematics, but is a general feature of science and engineering and elsewhere in our society. I see no solution. But in its incongruity lies an opportunity: if you are able to transcend it, you will be invaluable.

* * *

In Vietnam in June 2005 I met someone who has transcended it. David Mumford and I were the principal American invited speakers at an International Conference and Summer School held in Quy Nhon over two weeks. It was incredibly hot and humid, but David and his wife Jennifer and I elected to go on a three-day driving tour up to Hoi An, Hue, Da Nang, and My Son – all north of Quy Nhon. We also later traveled up to Hanoi together where I gave another lecture. All in all we had many interesting discussions and became friends.

David is a very famous mathematician within the pure mathematics community. He won the Fields medal for work in algebraic geometry and could be compared in brilliance to Alexander Grothendieck in France. Then, although not in the dramatic fashion of Grothendieck, David said, "Enough!" and moved from Harvard University to Brown University, and from pure mathematics to applied mathematics, especially the mathematics of vision. I have heard some of my pure colleagues castigate Mumford for his decision. But in my view, David had just decided it was more responsible to increase the breadth dimension of his mathematical contributions.

The first night in Vietnam we met at a hotel restaurant in Ho Chi Minh City (Saigon). David saw me eating alone and surmised that I must be the other American speaker, and came over to introduce himself and his wife. From that moment we never had any trouble at all communicating mathematically during our time together in Vietnam. He explained that one reason he became interested in vision was because he was colorblind. Our conversations could range from what is the space of all mappings of a circle, to my explaining the role of tension in deriving the wave equation. At a banquet, some Vietnamese musicians were playing various ancient stringed instruments, and David jumped up and insisted on trying to play them, much to the delight of our hosts. When the three of us were invited to a private dinner in Hanoi and the director of the institute wanted a photograph of himself with David, he insisted that I be included too, saying he did not believe in cults of the personality.

In my opinion, David Mumford made a better decision than did Alexander Grothendieck, if his choice was between changing mathematical persuasion and burnout. Of course there are undoubtedly myriad personal factors in each instance that one cannot know.

I share an inherent trait with David, and one that I conjecture is shared by many mathematicians. One evening when the three of us were on the main shopping street in the old city of Hoi An, Jennifer finally succumbed to the entreaties of a lovely young Vietnamese woman to enter a silk clothing shop. We had been in several already, but this shop had a special upstairs where one could see the silkworms and the whole process of unspinning their cocoons to form the silk fabrics. Clearly once one went up those stairs, it would be difficult to escape the further sales pitch. Jennifer wanted to go up. I still remember the grim look on David's face as he looked at me and saw that I too was totally paralyzed, as often happens to me in shopping situations...sometimes my sight and breathing almost fail. We math guys just go zombie when faced with shopping! But then David and I both simultaneously forced gigantic smiles and followed Jennifer up those steps.

* * *

In 2007, somewhat by chance the famed British mathematical physicist Sir Roger Penrose accepted the invitation of the Mathematics Department as our annual lecturer. Penrose was to be on a speaking tour touting his new book, *The Road to Reality*, and a week in Boulder fit perfectly within his schedule. Within our department, only I and one other could really engage in intelligent conversation with Roger about matters in quantum mechanics, relativity, and just about everything else in current theoretical physics. It was a great week and we had him essentially to ourselves within the Mathematics Department. The physicists on campus were amazed and frustrated, and turned out in the hundreds for Penrose's three lectures. But it was quite quiet on the Mathematics Department front.

I showed Roger my results in which most of the Bell-Wigner-Clauser-Horne-Shimony-Holt hidden variable theory inequalities may be embedded and understood geometrically within my operator trigonometry. Next our discussion turned to my rigorous investigations of the Zeno Paradox in quantum measurement theory, the so-called "a watched pot never boils," which in my opinion is fundamentally a mathematical rather than physical issue. Roger was quite receptive to my new Bell inequality trigonometry and suggested I should go further and try to see them as measures of entanglement. Recently I have done so. As to the Zeno paradox, our discussion soon focused on the claimed quantum phenomenon of wave function collapse. This contentious assumption says roughly that when you perform a quantum measurement, you collapse the incoming information onto the eigenfunctions of the measuring instrument. Penrose wants to bring in Gravitation as the missing ingredient. I would seek an explanation hidden within quantum mechanics itself. In my theory, if you rigorously follow the quantum evolution over time, I have proven that the Born probability picture (or "Born rule") is placed in doubt; for you cannot lose even a single prepared eligible probability state, not even at any instant in time, if you wish to assert any probabilistic conclusion about the other prepared states. Roger pondered that a bit, but fell back on the more conventional view that you do not have a probability until you actually perform a measurement.

Notwithstanding his age, about 75, and notwithstanding a snowstorm and plunging temperatures, Roger insisted that we take a hike in the Colorado high country the next day. Parking the car at a 10,000 foot solitary trailhead, he and I, and Professor Homer Ellis, put our heads down and headed into a blizzard, with temperatures hovering near zero degrees Farenheit. During this three-hour rather epic struggle against the elements, we discussed all the big issues of current theoretical physics. I found in Roger a kindred spirit with whom my own views largely coincided. In brief: we do not favor string theory, we believe in seeking classical descriptions in three-dimensional space, or if not, at least in low and realizable dimensions.

One of my favorite questions to top theoretical physicists is, "What is a field?" Everyone uses the concept of a *field*, for example the magnetic field, or the field theory of elementary particles. If I get a quick preemptive reply, my opinion of that person drops quickly. Not so with Penrose. We traced the idea all the way back to Newton. Roger pointed out to me that Maxwell actually devised a system of mechanical wheels to try to help him visualize a field. I pressed the issue and pointed out how Dirac didn't really believe in his quantum field theory. It has been almost a miracle how the introduction of fields as a mathematical artifice has yielded the fantastic differential equations of both classical and quantum mechanics. But the field still begs the physics.

Another thesis we share concerns the use of complex analytic function theory in both mathematics and physics. Penrose speaks in his book of the magic of complex numbers when used in physics. I maintain a somewhat original viewpoint, which I did not present to Roger, nor have I published it anywhere. It is that when you employ one of these analytic functions $f(z) = u(x,y) + iv(x,y)$ in any domain of mathematics or physics, you are actually insisting that the mathematics or the physics is of an incredibly self-averaging nature. By this I mean that each of the component harmonic functions u and v have that property, indeed may be characterized by that property, of necessity always being their own spherical averages. This is so well-known to all of us in partial differential equations. It gives to the analytic function f(z) therefore such an unavoidably averaged value physical or mathematical meaning. Moreover, the same applies to virtually all of the derivatives of f(z). So I want everyone who employs complex functions, be it in pure mathematical algebraic geometry or in some physicist's analytic continuation argument in some remote paper he is writing, to fully comprehend that they are forcing onto whatever objects they are treating what I would call "mean value mathematics" or "mean value physics." Because the use of complex numbers and complex analytic functions is so ubiquitous in advanced mathematics and theoretical physics, my thesis is that at every step of the way, the mathematician or physicist should pause and reflect: "Hmm...what does this self-averaging mean at this step?"

* * *

I have tried to indicate in this chapter the "flavor" of mathematics as it has tasted to me in my experience. As anyone knows, *chaqu'un à son goût*. To each of us, different flavors appeal. So another mathematician would write quite a different chapter.

My interests have been broad and often I am uncomfortable with how technically weak I can be at times. I rely much more on intuition than do most mathematicians. Sometimes I think my prowess has been due in large part to a rather large and active, even uncontrolled at times, imagination. In this respect I have identified with John Nash. Some years ago Jakob Bernasconi and I were involved in some problems in game theory, and out of curiosity went to Nash's web page. There was a paper on collaboration in game theory in which suddenly John went off into a discussion of fluid dynamics. I went right along with him, which greatly amused Jakob.

I teach in the same way, stressing the main ideas, the intuition, the large view, at the expense of training my students to be excellent technicians. That I must leave to those who are stronger technically than am I.

Notes

You may Google or otherwise access my webpage for more details about my mathematics within this journey.

Grothendieck's memoir is entitled, *Semaines et Recolte* (Seeds and Harvests), and emanated from the University of Montpelier.

The history of the French Bourbaki School may be found in *Bourbaki: A Secret Society of Mathematicians*, American Mathematical Society (2006).

Penrose's book, *The Road to Reality,* was published by Alfred Knopf, New York, 2005.

12. High Finance

...and dangers of overquantization...

"Infinite wealth!"

Schachermayer paused, chuckled, and beamed at this familiar gathering of financial derivatives friends who had gathered in Germany to honor the retirement of one of their own, Professor Hans Föllmer. We were in Berlin and it was June 9, 2007. I was there by chance.

Schachermayer looked again at the screen where he had projected his exponential wealth gain equation, grinned, and continued his explanation of the mathematics, then paused again, and in his relaxed and lively manner and with even a twinkle in his eye, repeated the mantra: "You can become exponentially rich. Moreover, here is how you can even invest that last penny!" The very small group of experts which constituted this elite assembly laughed, some in good humor, others a bit anxiously. I had learned more in three days than I could have learned in a lifetime of study.

In the previous lecture, Madame El Karoui, who was the "dean" of the financial derivatives community in France, as she started her presentation, had looked at her old friend and our guest-of-honor, the retiring Hans Föllmer, and chided Föllmer for being such an academic. And in splendid jest said, "Now, Hans, perhaps you will join the rest of us in lucrative consulting for the banks!" Föllmer just smiled, and quietly shook his head...no, he was not going to do that. Then El Karoui went on with a technical presentation of her group's attempt to price a strange but widely used derivative instrument called CDOs (Collateralized Debt Obligations). These, she said, were now in extensive use by the banking community and involved extremely large sums of money – in the billions. Instead of employing the more modern methods called Black–Scholes differential equations or stochastic processes martingales, she had fallen back onto an old multivariable statistics approach called Stein's method. It seemed to me that she would have a problem doing that with a large portfolio, and when I asked her after her talk, she amiably agreed, "You are right. We can't really price these portfolios. They have hundreds of underlyings in practice, and we can't get there...."

On August 10, 2007, the Dow-Jones Average fell 387 points and the credit panic of 2007 began. CDOs and similar financial instruments were at the heart of the collapse.

* * *

Why was I in Berlin for that select conference? The answer is a sequence of chance events; that sequence a nice illustration of the stochastic world through which our life trajectories proceed.

In 1988 I had helped obtain a $22 million NSF Optical Computing Systems grant for the University of Colorado. I was in charge of mathematical algorithms to be tested on the evolving hardware. A particular goal was to see if the high parallelism of optics could be married to the high complexity of neural networks to produce a new sixth generation computing paradigm, more capable of learning and even thinking than are the current fifth generation digital computers.

In the spring of 1989, on the street one day I met my colleague Walter Wyss, and was introduced to his Swiss friend Dr. Jakob Bernasconi. Casual conversation revealed that Jakob had also been plunged into this new science of neural networks by his employer Asea Brown Boveri (ABB). His task was that of investigating whether neural networks could model financial trading. As a consequence of this chance conversation, I would be invited to Baden, Switzerland, in the summer of 1989 and several times thereafter for joint research with Jakob.

In 1994 I wanted to spend one month of a sabbatical semester in Baden and Jakob said, "Fine. We will ask you to write us a survey of risk in portfolio management while you are here." ABB supplied me with J. P. Morgan's just-released Risk Metrics (1994) documentation. Also I embarked on a lot of background reading...and I could not believe my eyes when I saw it: There was one of the most important partial differential equations in the world, the now famous Black–Scholes equation,

$$\frac{\partial V}{\partial t} + \frac{\sigma^2}{2} S^2 \frac{\partial^2 V}{\partial S^2} + rS \frac{\partial V}{\partial S} - rV = 0$$

which is used to price financial futures and options. I was stunned. As the author of a book on partial differential equations, how could I not have seen this one? The answer is that most PDE books deal mainly with the principal PDEs of physics and engineering, and not of finance. I quote from my 1994 ABB report: "These particular partial differential equations are new to me, and as an expert in partial differential equations, I would be interested in getting to know them better, toward possibly improving their use for financial modeling purposes." In retrospect, this was a very naive understatement. But nonetheless, I was able to teach the Black–Scholes theory a couple of semesters and have recently produced two Ph.D. students in mathematical high finance.

In the Black–Scholes equation, for which the Nobel Prize in Economics was given in 1997, V is typically either the price of a Call or Put option, S is the stock price, r is the U.S. Treasury rate (assumed riskless...until recently), σ is the stock volatility, and t is time running forward. This is not the place to go into further detail and it should be emphasized that the theory of these financial derivatives has dramatically advanced in the last thirty years; so much so that I sometimes tell curious inquirers, "It makes the PDEs of physics pale by comparison." I managed to get a brief derivation and explanation of the Black–Scholes equation into the third edition of my PDE book, including how one can by a change of variables reduce it to the standard heat equation. It was quite remarkable and even frightening to realize that our financial markets were being run by the heat equation!

The validity of the use of the Black–Scholes equation rests upon an assumption of no-arbitrage, for which the roughly equivalent but more descriptive terminology "no free lunch" is sometimes used. The idea is that buyers and sellers are so active that whenever they place a bet in a belief of a free profit, it gets evened out by the opposing bets. The result is that would-be arbitragers all cancel each other out and a correct price is produced.

What really makes the use of financial derivatives so interesting is their coupling to the humans using them, and more to the point, to human nature. For example, it is not so widely known that a major reason for the Crash of 1987 was an aspect of what is called model risk. In that instance, too many of the *quants*, i.e., the mathematicians designing financial instruments, were using essentially the same Black–Scholes equations. So the needed balance of different opinions producing opposing bets was lost. The situation was exacerbated in 1986 when the CBOE (Chicago Board Options Exchange) put in a system called Autoquote which would take existing option prices from the market and generate synthetic stock prices to fill in the gaps in an incomplete market. Then traders would have a wider range of options prices from which to choose. This created a circular situation in which option prices drive both synthetic and real stock prices, which drive option prices. The Black–Scholes equations need to assume that the S in them, the stock price, is known to all who use the equations, and that S is not dependent upon V.

If you are a quant or a trader, no matter which, you must make as much money for your employer as your colleague does for his. So you are driven to use the same, latest technology. That is why in 1987 most of the firms on the CBOE were using essentially the same Black–Scholes models. The result was that instead of the models modeling the market, the market modeled the models! From 1976 until 1987 the Black–Scholes–Merton model was a close fit to observed market option prices. Then on Monday, October 19, 1987, American stocks took their largest one-day fall in history – more than 20 percent.

In game-theory terms, everyone is a free-rider on the way up, but when the crash comes, each individual bails to try to cut his losses, and the buyer-seller delicate balance from which the Black–Scholes equation is derived is no longer valid. I call this a *suddenly incomplete* market, a possibility that is always hiddenly present. After 1987 the Cox-Ross-Rubinstein improved version of the Black–Scholes model allowed the trader some flexibility, and also has built into it a probability of a crash. But no model can survive when all the buyers rush to the exits.

When a market drops precipitously as it did in 1987, large sums of collateral start to move to try to cover the obligations of market participants. Volatility goes up and that causes option values to shoot up. Market makers who had borrowed to sell their options saw those options shoot up in value and they could not make the huge margin payments owed to the clearing firms. Money ran out and the markets collapsed.

So you need arbitragers – the more, the better, it seems. Most large financial firms have their arbitrage desks. How finely you can arbitrage is essentially limited by transaction costs.

The CU Business School sent me enough Ph.D. finance students to combine with a few Mathematics graduate students, allowing me to meet the required enrollment numbers and teach the Black–Scholes–Merton theory in the spring semesters of 1999 and 2000. From those courses emerged my first Mathematics Ph.D. in Finance, John Davenport. We looked carefully at the arbitrage that can occur when the U.S. Treasury interest rate r in the Black–Scholes model becomes slightly out of sync with the actual underlying stock price drift. In particular we showed that for the American Put option, which is a right to sell an asset at any time up to the option expiring date, the existence of an impending dividend payment can cause the so-called exercise boundary to fold over on itself, thereby branching with respect to the time variable. In particular this result revealed a critical weakness in what is called the Geske-Roll-Whaley dividend model.

My second mathematical finance Ph.D. student Troy Seguin approached me in 2004 with the intriguing and quite challenging proposal: "Let's price Venture Capital!" Such an idea is audacious in view of the very clear incompleteness of Venture Capital markets. One needs an assumption of complete markets in conventional financial instruments pricing theories. We did not succeed in entering the Venture Capital world, but we learned a lot. In particular, we were led to study recent attempts to create a theory of risk that is general enough to cover both incomplete market risks and also non-market risks. This led us to the 2004 Föllmer–Schied book *Stochastic Finance*, in which such recent research on the development of general risk axioms is treated.

The chain of chance events taking me to the Föllmer retirement colloquium is not quite finished. I had been invited to speak at a quantum mechanics conference in Växjö, Sweden, June 11–16, 2007. While internet surfing a couple of weeks before I was to depart, somehow a random click brought a small announcement of the Föllmer colloquium to the screen. A quick email to the organizers gained their permission to attend. So to help my Ph.D. student, I adjusted my travel plans to leave a few days earlier, allowing me to attend the Föllmer meeting in Berlin.

* * *

The "infinite wealth equation" Walter Schachermayer was tantalizing us with was an exponential expression of the type you will find in the Föllmer–Schied book. There, instead of just using stochastic differential equations like Black–Scholes, you may also use the martingale (fair game betting) theory of stochastic processes. Then to price a financial instrument you must introduce so-called Utility functions, which, roughly, are presumed to balance off how much risk you are willing to entertain in order to make a certain profit. Once this mathematical machinery is in place, then you attempt to optimize your gain under the risk constraints.

Of course, you may also lose. So you set a loss upper bound X which is to represent the maximum amount you are willing to lose. As my student and I learned this theory, we kept coming back to an uncomfortable feeling that this number X was like a tail wagging the dog. Every time you did some clever stochastic mathematics to optimize your wealth ever upward, there seemed to be a need, either explicit or implicit, to also fudge your maximum loss X upward. One way that this increase in potential loss X is present although not evident is hidden in the fact that you never really know what is called the Girsanov equilibrium probability measure Q. That's because you never really know the actual true probability measure P from which you must transform to the Girsanov measure before you bet.

I quickly surmised in Berlin that I had fortuned onto a select community of the world's best quantitative finance experts. From my already ten years of reading and working with my students in the subject, I could take full advantage. One often can learn much more from a casual conversation than from even the best of lectures. Were banks really using these mathematics? Some of the younger academics did not know, but I was assured by a couple of the elder statesmen that, yes, indeed so, the banks are making big money and betting heavily, using the martingale approach. How are they implementing it? Monte Carlo. That is a technique where you run a large sample of situations on the computer and afterward average over those runs to make a prediction of your gain or loss. There is a key presumption that your experimental ensembles adequately cover the probability event space. We now know of the great danger, or in modern parlance, that you have likely not taken into account any Black Swan Event (an unpredictable event of large consequence which lies outside your model).

So in a nutshell: if you want to try for infinite wealth, there is probably a hidden possibility of infinite loss. Those hidden possibilities of huge losses can be obscured by the sheer complexity of the financial instruments. Moreover, there is almost no way to monitor and regulate all the transactions between parties which in effect link all these instruments together. Thus we have here a large number of unregulated fast-acting, highly-interlaced complex financial trajectories, carrying huge sums of money in an event probability space whose risk measures we do not, and probably cannot, adequately model.

Only two months after the Berlin colloquium, the credit markets began to collapse. There has been, in hindsight, of course much analysis of the resulting general financial crash, the worst since the Great Depression. Congressional hearings placed blame on the bankers, but not on themselves for having removed the needed regulation enacted after the last Great Depression. I followed all of this avidly, even as my own retirement funds shrank by a third. It was quite fascinating to watch politicians grappling with financial engineering without understanding its underlying mathematics.

I knew all along, let me call this my first assertion, that it was essentially a matter of human greed in placing a value of X too high in order to place a higher-yield bet so as to get a higher trade commission. A commission in the millions...very tempting! Others who read this may of course have their preferred scenarios. Fine. Let us place all those scenarios into what I would call a created hypothetical scenario-market, out of which should emerge a correct combined scenario – in analogy to the options and credit markets. My second assertion is that all markets, even our new hypothesized scenario market, are incomplete. Proof: add your own scenario.

When collateral sums are insufficient, a market collapses. That is what happened in the 2007 credit market panic. And those CDOs that Madame le Professeur El Karouri admitted to me could not be priced, were front and center in the crash. Generally, CDOs buy a portfolio of fixed income assets, such as sub-prime house mortgages and commercial mortgage-backed securities. These CDOs were traded between banks and other financial institutions. For example, often a bank would buy a AAA-rated CDO and simultaneously purchase as insurance a CDS (credit default swap). So the risk of defaults by over-leveraged home owners was transferred to the CDO buyers who then transferred the risk to the credit default swap writers.

Huge commissions were made on these deals because large sums of money were inherent within these financial instruments. It may seem a bit coarse, but when I explain it in non-financial terms in casual conversation, I just say, "Hey, I will loan you a billion, at a 10% commission, then in a few weeks, why don't you loan me back a billion, at a 10% commission?" With no industry-wide rules or overall regulation, a $33 trillion underbelly of hidden systemic risk to the global financial system quickly developed.

What is extremely ironic is that all of these extremely risky bets could have continued, had there been no market. In other words, even if we could not mathematically accurately price these massive financial instruments, the game could have gone on, at least for awhile, even as the housing prices started to decline. But in 2006 new indices, so-called ABX indices, appeared in the over-the-counter market. These indices allowed market participants to buy or sell protection against defaults of sub-prime bonds and mortgages. Thus, even though we could not mathematically price a CDO, and in fact parts of its portfolio were informationally out of reach, nonetheless, the humans could now financially express their opinions about the risks. As a result, in 2007 the bottom fell out of the ABX market. The Sub-prime crisis of confidence had begun. Big institutions like Goldman-Sachs rushed to get their collateral calls first in line. Market-to-market accounting rules then accelerated the price declines in a multiplier effect. Large scale illiquidity by large financial institutions followed. Rather than letting the whole world financial system collapse, American and European governments stepped in with their massive financial bail-outs.

But those who constructed and issued the CDOs had already pocketed millions or billions in trade commissions. They really had no incentive to truly price the risk. They effectively took the money and ran, leaving unsuspecting investors holding the bag. These investors then ran to the government for help. As one bumper-sticker put it: "Capitalize the Profit, Socialize the Risk."

* * *

It is really quite remarkable that I had stumbled into the role of well-informed and well-trained outside observer of one of the greatest wealth reduction epics in world financial history. My student Troy and I had started out to price venture capital, a notoriously incomplete market. Instead we were witness to the collapse of the huge sub-prime market as it was revealed to be also fatally incomplete.

* * *

About ten years ago, a serious effort at dealing with incomplete market risk was initiated by a number of researchers. The goal was to somehow accommodate both model risk and human risk factors. The result was the development of what are called convex and coherent risk measures. An account of some of these may be found in the Föllmer and Schied book (2004). Troy and I studied them further in his Ph.D. thesis (2008). Without going into the technical assumptions underlying them, it is surprising that to achieve generality in applicability, these risk measures need only satisfy four axioms:

Subadditivity: $\rho(X + Y) \leq \rho(X) + \rho(Y)$
Homogeneity: $\rho(\lambda X) = \lambda \rho(X)$
Monotonicity: if $X \leq Y$, then $\rho(X) \geq \rho(Y)$
Translation: $\rho(Y + m) = \rho(Y) - m$

To fix ideas, think of X and Y as two portfolios of investments. Then, roughly, subadditivity says that larger diversification reduces risk, homogeneity says that increasing the amount of a fixed portfolio increases risk, monotonicity says that a better portfolio will carry less risk, and translation says that if you add riskless assets to your portfolio, you decrease risk.

Despite the sparseness of these four axioms, a lot of technical mathematical machinery for their implementation into investing strategies and organizational risk management has been developed. For example, your own attitude toward risk can be encapsulated into a penalty function which in conjunction with the axioms and your portfolio asset values will tell you what your own maximum loss X mentioned above should be. Of course, less maximum loss X means less potential profit.

If you are investing other people's money, as was the case with bankers and traders dealing with CDOs and CDSs, you can potentially gain infinite wealth, or at least a disgustingly obscene large effectively-infinite-by-a-laborer's-standard commission, by setting X almost infinitely large.

* * *

I have found this field of high-finance mathematics fascinating. It combines advanced partial differential equations, with advanced probability theory, with advanced financial and economic theory, with advanced human psychology, with outrageously large sums of money. Sometimes I call it Financial Fysiks. Warren Buffett famously called derivatives "financial weapons of mass destruction." Yet he, and large parts of the world economy, are as locked into the use of financial derivative instruments as we all are to home and auto and health insurance. Farmers weather bad years and gold mines survive commodity price swings by being optioned. So the problem is not the mathematicians, the quants, who are called upon to design these instruments and understand this fantastic financial fysiks. Just as it was with the scientists who were called upon to develop and understand atomic bombs, the use of advanced mathematics in finance is inevitable. The dangers lay in its potential abuse by human greed coupled with human ignorance. How to limit that abuse by humans is the challenge our politicians are currently (and in many respects, unsuccessfully) grappling with.

* * *

As for me, I remain largely an outside observer. I do have a number of interesting ideas. Of course, I do not know what to do with them. Here are some:

In teaching the Black–Scholes partial differential equation theory in the late 1990s, I was struck with the fact that the mathematics treats the stock price S and time t as two independent variables over which you plot the option price V(S,t). This is Cartesian thinking. Everyone knows that stock price S and time t are not really independent in financial markets. However, the Black–Scholes theory works because it is derived from the assumption that the stock price changes ΔS are random. What bothered me was the underlying issue of whether the stochastic independence of causally arriving random price change increments was consistent with the Cartesian independence of algebraically and functionally complete independent variables S and t.

Accordingly, I assumed not, and wrote a paper called *A 9th Derivation of the Black–Scholes Equation* and presented it at a conference on Computational Methods in Decision Making and Finance in Neuchatel, Switzerland in August, 2000. But the referees did not like it or did not understand it and rejected it for publication. Remember that the original Black–Scholes paper was twice rejected for publication! I went ahead and presented the basic ideas in a short section in my chapter *Time-Space Dilations and Stochastic-Deterministic Dynamics* in the book *Between Chance and Choice,* Imprint Academic (2002). I concluded that the very notion of hedging (instantaneously) is a myth. I speculated that this myth might be obviated in practice by the rapidity of updated re-pricing. Now I wonder if there is a connection to the recent practices of flash and naked trading wherein whether a computer-driven high-frequency trade will win or lose depends on being a few hundred microseconds ahead of the other fellows. Such practices, variations on front-running, are reputed to have made Goldman-Sachs their record multibillion dollar profits this year.

Another is a set of ideas I have about that fourth axiom in the risk measures. I am unable to extemporize those ideas fully here. But there is an underlying notion of cash-invariance there, which roughly allows you to move cash into and out of your portfolio without consequence. Perhaps I will further develop those ideas, with or without a new Ph.D. student.

Then there is the confusion between far tail events and black-swan events. The former can be expected, albeit rarely, whereas the latter cannot. Why can't the latter be modeled? Because they correspond to events that lay in the future and also perhaps beyond our imagination. So our working event-algebras have not anticipated them. Therefore we need a new probability theory. I have thoughts on that, but again cannot elaborate them here.

I also find confusion in the finance literature between far tail events, fat tail events, and rare events. Far tail events can be accommodated and are the basis of the Basel Bank of International Settlements Value at Risk formula for imposing risk limits. The dangers of its limitations are well understood but are counterbalanced by its practical implementability. Fat tail events can be adequately modeled by using statistical distributions having also a fourth moment, and I even did this in 1960 for bearing angles in my secret work in direction finding for the U.S. Navy. Rare events usually can be treated by known statistical theory for them. But all of these mathematical models presume some underlying *a priori* statistical distribution and moreover that it may be treated by continuous calculus methodology.

Complexity theory and its power laws and links to chance and connectivity have become very popular recently with financial and management gurus. I have read a number of their books and I am usually somewhat disappointed. As one who has worked mathematically with nonlinearity and so-called chaos for many years, I get the feeling that these gentlemen are jumping on the bandwagon in a perhaps earnest but doomed attempt to understand things that are beyond them. One recent book extols the sand pile model as exemplifying how a sudden massive shift may occur. The Greeks knew about the sand pile and its instability. And one does not need nonlinearity to have great instability with respect to initial conditions. We in PDE know the linear examples of Hadamard of one hundred years ago. Then there is the trumpeting of asymmetric warfare in which your terrorist enemy always stays just a half-step ahead of you. But that idea is already implicit in Charles Darwin's Red Queen hypothesis.

You cannot think outside of the box unless you already have lived and survived outside the box for some time, and know that terrain.

It seems that my imagination, going way back to my school days, has often taken me out of the box, often to the discomfort of others, sometimes much to my own discomfort, when faced with the forces of conformity.

Some say you should never trust correlation data. An example is the now infamous copula model that was used to price billions of dollars of CDOs before the credit collapse of 2007 but whose time-limited past data may have been misleading. But I have a quite different view, which is grounded more in thinking about dynamic ever-changing event spaces which also take into account human behaviors and tendencies. All correlations are indeed present but are always changing and with new innovations at all times. Therefore risk is non-Bayesian and irreversible. Some ingredients of this theory may be found in the discussions in the next chapter. Although there the emphasis will be on a theory of human trajectories, I would love to be able to also apply it to enable better understanding of the behaviors of financial markets.

In a recent memoriam paper dedicated to the late Professor Ilya Prigogine, I decided to dig deep into a somewhat mystical *principle of detailed balance*, which I had first happened upon in my work in nonlinear fluid dynamics. Later I had found that I could identify it as an essential step taken by some Nobel Prize winners in chemistry and physics. In particular, at a critical moment in their thinking, I found that they had put in by hand such a detailed balance to achieve their goals. In my analysis, they had, perhaps tacitly, perhaps implicitly, willfully or not, transformed a key microscopic physical dynamics which was at its heart irreversible, into something which was artificially rendered reversible and thereby conserving of key physical entities. Even Einstein did this to obtain his probabilities for the absorption of a light quantum.

From this I have come to a view of the Black–Scholes equation as a forced detailed balance of risk. I also see the complete market hypothesis as a forced detailed balance of buyers and sellers. I see the little picture of a gap between a buy price and a sell price as one of great micro-activity whose exact dynamics is always irreversible even though theory says it always closes. When I explain this idea, I put my forefinger down close to my thumb and then let them set up a very rapid vibrating, tapping motion.

I wager that this new field of mathematical science, which I could call financial-economic-psychological-societal mathematics, is just in it infancy. It is too important to be otherwise.

Notes

My ABB report *Risk in Portfolio Management* (1994) was internal and not published. The derivation and conversion of the Black–Scholes equation to the heat equation may be found on pages 431–434 in my book *Partial Differential Equations*, Dover, 1999. The Föllmer–Schied book is *Stochastic Finance*, 2nd Edition, de Gruyter, Berlin, 2004. My Ph.D. students' dissertations are *Analysis of American Options* (2007) by John Davenport and *Risk Measures* (2008) by Troy Seguin. Nassim Taleb's best-selling 2007 book *The Black Swan,* Random House, N.Y., eerily mirrors the Subprime event that was occurring simultaneously. The book *Between Chance and Choice,* edited by Harald Atmanspacher and Robert Bishop, Imprint Academic, U.K., 2002, is a collection of interdisciplinary perspectives on determinism. The memoriam paper I wrote for Ilya Prigogine is *Microscopic Irreversibility*, Discrete Dynamics in Nature and Society, 2004: 1 (2004), 155–168.

13. The Improbabilities

...and the imponderables...

We all seek to know the meaning of our existence. Why are we here to work, love, play, suffer, wonder, die? This question, the meaning of it all, is the great imponderable.

This tale of my life has revealed a trajectory interrupted at critical moments by unpredictable events. At those junctures the path taken has often depended on the close human relationships present in my life. At other times the fork of the path turned absolutely on a life-and-death chance event of the instant. I am led unavoidably to conclude that uncertainty plays a critical and ever-present role in our lives. And that conclusion gives rise to two central questions that I continually ponder and explore: (1) What is the nature of this sea of uncertainty within which our lives take place? (2) And, is this sea of uncertainty related to the meaning of it all, that great imponderable question?

I purposely use the unconventional term *improbabilities* in the title of this chapter to connote the first question, for I see none of our conventional probability theories capable of answering it. If I am right, that then makes the second question – which I term an *imponderable* – all the more imponderable.

I propose to examine the first question by placing the trajectories of our lives into a frame of three elements: *cause, chance,* and *choice*. To carry this discussion forward such that it engenders not only the first question above but also the second, I propose an ansatz – a working hypothesis – that God, taken here to be the whole universe, also has access to the three elements of *cause* (deterministic), *chance* (stochastic), and *choice* (by God, by humans). At each instant, God may allow some parts of the universe to evolve in causative deterministic ways, other parts to evolve by chance, and in some special cases, by specific choice.

To fix ideas, consider the following precise example from Chapter 1, "The Child in Iowa" of this book: Recall that child in the back of a Buick heading out of Iowa toward Colorado, a major and improbable change to my life. The cause might have been a chance affair of my mother which resulted in my father's choice to sell his successful business, in hopes of building a new life elsewhere. All three trajectory elements are present. As to the second question, one may even ask, did God have anything to do with it?

My beloved Aunt Leona would answer without hesitation. A fervent Baptist, she would declare, "Everything has a purpose." Aunt Leona had no doubts at all about a deterministic universe. She always told me: "The Lord has a purpose for everything. Everything will work out the way it was supposed to. There is a reason for all things which happen to us, good and bad." So for Aunt Leona, the answer to the first question was all: *cause*. Her answer to the second question would be all: *choice*. God's choice. Always. So both questions would be answered simply, definitively, and *cause* = *choice* (God's). Thus, absolutely no *chance* exists in this model. Nice. But it's too simplistic for me.

It was during Aunt Leona's visit when my family and I were in Switzerland in the summer of 1972 that an incredible coincidence occurred. That prompted me to wonder whether one could devise an important new probability theory that could help us better understand our lives. Then, upon my father's visit later that summer, another extraordinary incident of unlikely "trajectory crossing" reinforced my curiosity. Immediately, I postulated and wanted a probabilistic theory of the dynamics of all human trajectories. I wrote down a few rudiments of such a theory, but I was unable to push it into any kind of comprehensive mathematical form. For want of a better name, I termed it Trajectory Analysis. I believe this could be of fundamental importance and in particular could enable a better understanding of the two questions of this chapter. Although it might not answer the second question, I consider it a possible means of modeling the dynamics implicit in the first question. But completely aside from the two questions, I believe it carries its own independent mathematical interest.

In brief, here are the two incidents which caught my attention that summer of 1972 in Geneva, Switzerland.

Aunt Leona was on a sentimental journey through Europe to visit the sites she and her late husband, a U.S. Army Infantry colonel, had shared together after World War II. They had met when both were serving in the post-war occupation forces in West Germany. She was on her way down from Germany to visit Omaha Beach at Normandy, where he had been among the D-Day landing forces.

We were living in a small apartment at 4 rue de Beaumont in an older residential area of Geneva. I arranged weekend lodging for Aunt Leona at a nearby palatial old hotel, Hotel de la Residence, on Route de Florissant. On Sunday morning I joined her for breakfast, and there was only one other person, an older woman, having breakfast at another table. Eventually she came over and introduced herself as Portuguese and widowed. Her husband had been a Portuguese military attaché and they had spent two years in the United States during the war. "Where?" Leona asked. "Santa Fe," the woman replied. "Why, I was based there too!" Leona happily exclaimed, explaining that she had managed the Army Service Club there. A frown came over the old lady's face. She asked if Leona had happened to have known a girl named Dee Dee, who worked there. "Why, yes, she and I were roommates!" answered Leona. There was a long pause. Then the elderly Portuguese woman said flatly: "Dee Dee had an affair with my husband. It broke our marriage. We stayed together but only on a formal basis." Leona's face was ashen. We sat silently – Leona said she knew Dee Dee was attractive and had had admirers, but she never dreamed she would be involved with a married man. The breakfast was over.

Later that summer my father came over to visit us. He had never been to Europe. To help entertain him we reserved a beach hotel for a week in Ravenna, Italy. From there we went over to Florence for two days to see the usual tourist sites such as Ponte Vechio and the David statue. My scientific host in Geneva, Professor Jauch, had told me about an out-of-the-way and at that time hardly known (though much better known today) little museum called Istitute e Museo di Storia della Scienza. There we would see many of Galileo's original instruments that he had constructed and used to convince his funders and protectors. We found the back street and went up the stairs to the second level, paid the fee, and started wandering around among these fantastic exhibits: for example, a wooden ramp down which a marble would roll, ringing bells at appropriate intervals to demonstrate the constant acceleration of gravity. We lost track of time as we delighted in all these older scientific instruments collected by the Medici family over the ages. We had thought we were the only visitors but eventually we were approached by the only other visitor, a pleasant looking young American in his mid-twenties. "Where are you from?" he asked. "Colorado." "Really! I was in school there," said the young American. "Well, I'm a professor of mathematics at the university there," I said. Pause. Then he asked, "Do you know Professor Norton?" "Of course, he was in our department," I replied. Another pause, then the young man announced: "He flunked me. My only F. In fact, it was my only bad grade. But because of him, I could not get into medical school." He went on to explain that he had since taken other courses at another university to get his grades up, and was finally admitted to medical school, which he would begin in the fall. But his F had cost him three years. As it turns out, Professor Norton had not been awarded tenure at Colorado partially due to faculty criticism of his uncompromising habit of too-tough grading.

Coincidences abound. And in the summer of 1972 I of course knew the usual mathematical explanations. The classic example is The Birthday Problem. How many people do you need in a room to have a better-than-even chance that two of them will have the same birthday? Let yourself be the first to enter. Then the next person enters. Assuming birthdays are equidistributed over a 365 day year (they are not, incidentally), the probability that next person has a different birthday is 364/365. The third person enters and the odds against a common birthday are reduced to 364/365 times 363/365. When the twenty-third person enters the room the product of these fractions drops below 50% to

$$\frac{364}{365} \frac{363}{365} \frac{362}{365} \frac{343}{365} \cong 49.3\%$$

and it is more likely than not that there is a pair with a common birthday. Another well-known example is sometimes called Littlewood's Law, or the Law of Very Large Numbers. Assuming that a human experiences life at a rate of one event per second and is alert only eight hours per day, nonetheless within thirty-five days he will have experienced more than a million events, some of which will seem almost miraculous.

So unlikely events will happen to you; but you cannot know in advance which ones, or when they will occur. The Birthday Problem explanation just uses pure mathematical combinatorics; Littlewood's Law of Very Large Numbers just depends on humans being exposed to a very large number of events. What I wanted in my 1972 Trajectory Analysis was to find a deeper explanation, which would include the role and dynamics of our human activities. Do these activities create a special selection of human trajectory crossings? Do we humans create a *preferred universe* within the physical universe? I was sure that we do. And the resultant probability theory of such a human trajectory theory would be much richer than that of just combinatorics or very large numbers. It would be a new probability theory. But how to render it into a precise mathematical formulation? Therein lies the hitch...

It was easy to write down, if only in a subjective way, some of the basic ingredients of such a new theory of human trajectories. Obviously the fact that all of our trajectories were constrained to the surface of a sphere (the Earth) enhanced the probability of random crossings. Then culture enters, for example, the overhearing of the English language being spoken in Switzerland and in Italy, in the above two coincidences. Time of course is of the essence in any dynamical stochastic theory. For example, my Aunt Leona and the Portuguese woman were of the same generation, although that does not help to explain the specific event timing. I did not imagine, and did not want, just some completely stochastic theory of random dynamics. So to create a *causative* component, I imagined each human to have a world-line described by their own differential equation. In 1972 those would constitute about 5 billion differential equations, some strongly coupled by family, work, and the like. Then *chance* could enter to couple some of these trajectories according to behavioral correlations as well as due to some other stochastic variables and of course also just due to some randomness. And how would *choice* enter? I had chosen Geneva seven years earlier as a postdoctoral site. But wait – I had not chosen it, it had been recommended as a possible site, contingent upon my winning a postdoctoral award competition. Still, I had chosen to apply. Thus Aunt Leona's trajectory would potentially pass through Geneva. The Portuguese woman had chosen Geneva for an entirely different set of chance and causative reasons. I wrote down in a similar way a number of causative, chance, and choice predecessors leading to my father's visit to us in Geneva and to that coincidence in that particular museum in Florence.

However, to be true to reality, I concluded that each of our own trajectories has an unlimited number of previous trajectory links to other trajectories. So whether countable or uncountable, realized or not, our past is just too rich to know. We can arrive at finite lists only due to truncation caused by our limited memories and perceptions. So rich was this potential new Trajectory Analysis that I wanted in 1972 that it overwhelmed me. I just put it on the shelf and went back to more tractable work such as quantum scattering theory! Imagine: as you read these lines, you have linked your trajectory to mine. Even virtual, nonphysical trajectories are intertwined.

However, unknown to me, less ambitious theories had already started to develop. I was surprised when a colleague in Brussels, with whom I had casually discussed my wanted Trajectory Analysis theory, brought me a copy of the International Herald Tribune on June 22, 1998, which featured a large splashy article about the small world phenomenon, in which all humans are seen to be connected by at most six degrees of separation. I was frankly disappointed to learn that the small world phenomenon had been a subject of study by sociologists going back to the 1930s, had been considered mathematically as far back as 1957, and had now been jumped on by mathematicians *en masse*. In the dozen years between that article in 1998 and the present, a whole new science of networks has exploded upon us, including for example the study of power laws, clustering, and hubs. As an example of hubs, I happen to have low Erdos number 2: that is, I am linked to Paul Erdos through only one intermediary mathematical publication. That is because, by chance Frank Harary visited Boulder, and by chance I wrote a paper with him, and he had already written a paper with Erdos.

Erdos was a Hungarian mathematician who devoted his whole life to living and traveling with his mother, and writing more papers than anyone else. Erdos with his 1,500 published papers, and Harary with his 700, are the two most connected hubs in the research mathematics community. The internet has been graphed and shown to have nineteen degrees of separation. A whole scientific industry called Complexity Theory now exists.

But in my opinion, the theory I wanted, especially with its human trajectories each equipped with its three elements of *cause, chance*, and *choice* riding along on it, has not yet been achieved. Perhaps Google has come the closest. When its PageRank algorithm – which runs its search engine – came into being in 2000, the interconnectedness of human society underwent an order of magnitude increase. In 2005 I was excited to hear of the PageRank algorithm and accordingly learned and taught it in my graduate linear algebra course the next year, to see to what extent it may implement my earlier ideas of Trajectory Analysis. Some of my ingredients are indeed there. Let me explain how.

In short, the Google matrix is now of size multibillion by multibillion, and in principle its matrix elements consist of the relative importance probabilities of linkage between all web pages or other internet possible destination sites. To describe it simply, let H denote that huge n × n matrix, with its elements h_{ij} summing over columns to one, making H what is called a row stochastic matrix. But some sites may not point to others, so to assure that all sites are potentially reachable, rows in H with all zeros are replaced with the uniform random probabilities 1/n, written mathematically as a long n-vector $e^T/n = (1/n, 1/n, \cdots, 1/n)$. Call this filled-in and now strictly positive row-stochastic matrix S. But the web may not be strongly connected in the sense that it may cluster around hubs and you may become trapped within one of the clusters. So to prevent that, S is replaced with a weighted average of itself and a uniform link matrix E, where E is just the large n × n matrix with all elements having value 1/n. Such E is conveniently written mathematically as ee^T/n, where e just denotes the vector of all ones. The weight coefficient α is found empirically and is about 0.85. Thus the Google matrix has become $G = \alpha S + (1 - \alpha)E$. Later Google replaced E with a more general ev^T where v is a vector which may be personalized to your webpage. How well does the resulting Google matrix

$$G = \alpha S + (1 - \alpha)ev^T$$

implement the ingredients of my earlier Trajectory Analysis?

Cause is represented by the deterministic Bayesian past relative importance of or by the current relative number of pages pointing to your page that gave the original matrix H. Granted, this is a stochastically based cause, but let us accept it as the deterministic element. *Chance* is represented by the random added rows and by the random noise matrix E added to stochastically link all sites. *Choice* has come in with the later added personalization vectors v. The time-dynamics I sought are there in discrete fashion when the whole Google matrix is updated each month. And they are there in each personalization vector which has, among other things, one's search history.

The Google story is also fascinating to me because in the late 1990s I thought to apply my operator trigonometry of operator turning angles and antieigenvectors to the earlier Latent Semantic search engines. In those, the relative importance of search attributes was measured by angles between vectors, so I could immediately connect them to my theory. However, I could not produce anything dramatically new so I dropped the project. What is then doubly fascinating to me is that the PageRank algorithm calculates the most important eigenvector of the Google matrix, and that determines those pages of link responses to your query. So I could, stretching and in terms of my own personal intuition, say that the antieigenvectors of Latent Semantic search lost out to the first eigenvector of Google search. That is because the founders of Google returned, due to the huge size of their matrix, to the most primitive method to calculate their first eigenvector, the so-called power method. That finds the principal first (left) eigenvector of G, whose largest components give those first 10 links on a query response page.

In retrospect, it is hardly surprising that I could not cast my Trajectory Analysis into a tractable mathematical frame in 1972. The personal computer had yet to be invented, the satellite communication system was not yet widespread, and there was no internet, all key factors in Google's success.

Nonetheless, Google is a robot. It cannot provide many of the goals of my theory. What correlates my not falling from the Window route in 1953 with my seeing a mathematical paper by Ed Nelson in the Battelle library in Geneva in 1965...and from which my first publication caught the eye of John Nash, who would win the Nobel Prize in Economics in 1994? The collateral event space needed for all the activities of each of me, Nelson, and Nash, is far too massive for any robot to search. Yet the trajectories of just the three of us tell the whole story. What explains how as an engineering student I witnessed a new satellite called Sputnik go overhead in 1957, causing me to be called into teaching mathematics, and within a few short years being thrust deep into critical military espionage and the world's first spy satellite...and software used in the Cuban Missile crisis which had put the world at the edge of nuclear war? Each run of Google's search is actually a deterministic one and cannot take into account the chance and randomness underlying our trajectories. The associated probabilistic event spaces are of course forbidding in size and I would assert even unavailable.

But it is much easier to see my trajectory as being carried along, bunched with the trajectories of other individuals of my age and training, into a general flow of

Cold War work. This provides a strong and simplifying correlation. For another example, how would a search engine predict that a child in Iowa would collaborate with a physicist who did not obtain tenure in Boulder, but who would link me to a Nobel prize winner in chemistry in Belgium? The related event spaces here seem even more disparate; but just knowing our three individual trajectories would seem to render meaning. We don't even need a correlation such as the Cold War work: We just need the chains of events of our three lives.

Therefore let us look more closely at what my three desired elements of *cause*, *chance*, and *choice* may really intend. The event space of *chance* may be too large to contemplate, and *cause* if not known too hard to establish, but *choice* guides trajectories. In research on neural networks toward financial trader analysis, Jakob Bernasconi and I became sidetracked on how important *context* was. I quote from the abstract of a paper we published in 1998. "A major implication of our findings is that humans overwhelmingly seek, create, or imagine context in order to provide meaning when presented with abstract or apparently incomplete or contradictory or otherwise untenable situations." In a related 1998 paper I distinguish *context* as more general than *agenda*, the latter being a plan of procedure, the former a setting and its meaning. Our trajectories are an encapsulation of our goals and our agendas to attain them and the contexts within which we may do so. *Choice* sets goals and agendas and selects available contexts and thereby is an important factor on where our trajectories will reside in time and space.

For example, the essentially independent event spaces of Ed Nelson, John Nash, and me are nonetheless clearly linked by our common choice to be mathematicians, from which the first example I gave above becomes quite reasonable.

Looking now at *cause*, let us take advantage of the second example above. To repeat, although there were many choices involved, nonetheless the powerful cause of serving the country in the Cold War dominated our lives and hence their trajectories in that period. As to *chance*, well, I see it as ever-presently capable of affecting all trajectories. I would therefore attribute the third example above leading to my meeting Ilya Prigogine as mostly a matter of chance, albeit explainable in hindsight by trajectory crossings.

Generally, we humans prefer to have an explanation via cause rather than just settling for chance. We have figures of speech such as "in any event" or "whatever the cause," which we unconsciously employ in our conversations because we cannot settle the cause.

* * *

As part of our continuing research on the physics of irreversible processes, in 1993 Ioannis Antoniou and I published a paper entitled *From probabilistic descriptions to deterministic dynamics*. This was a technical work in which we proved that we could mathematically extend a probabilistic Markov process to a larger deterministic Kolmogorov dynamical system of which the stochastic Markov process was a projection. The larger system was still chaotic and ergodic but was, in principle, predictable. In short, one could interpret our result as that of taking a chance process and rendering it into a causative, deterministic process.

We were astonished to receive a reprint request from the R. J. Reynolds Tobacco Company. One must recall the long and expensive legal battle between the United States Government and the tobacco companies to recover from the latter the massive health costs to the government, purportedly due to humans smoking cigarettes. This battle was a stalemate until scientists were finally able to establish beyond a doubt in their laboratories *the causative mechanism* through which the cigarette chemicals induced cancer. Prior to that, all the established strong statistical correlations of tobacco smoking to cancer had been legally insufficient. Just today, 2-20-10, I note by chance the following news item in our local newspaper: "Obama administration asked the Supreme Court Friday to allow the government to seek nearly $300 billion from the tobacco industry for a half-century of deception that 'has cost the lives and damaged the health of untold millions of American.'"

An analogous controversy rages within the physics community concerning the foundations of quantum mechanics. The issue is aptly summarized in Einstein's statement: "It is hard to sneak a look at God's cards. But that he would choose to play dice with the world...is something that I cannot believe for a single moment." Einstein wanted *cause*. Niels Bohr is reputed to have told Einstein: "Stop telling God what to do!" Bohr had accepted *chance*. My theory's third ingredient, *choice*, would seem for its existence on the quantum level to necessitate that Bohr was right. I have written a number of mathematical papers on the quantum measurement problem: whether the chance element really depends on our act of measuring it. Does choice really enter fundamental physics on an equal footing with *cause* and *chance*? On the larger macroscopic level, *choice* is often associated with the issue of how much free will we may have in our lives.

* * *

Thus we have been led to the second question of this chapter – that of whether the underlying uncertainties, be they at the microscopic or the macroscopic level, are related to the meaning, or meanings, of our lives. Certainly each of our life trajectories is greatly affected by the macroscopic uncertainties which we cannot predict and cannot avoid. Quantum mechanics has forced uncertainty down to every atom, including those of our own bodies. Are these uncertainties, be they by design or by accident, a fundamental part of the meaning of our lives?

In raising the second question, this chapter has now arrived at a point of great instability. Religious authorities and philosophers and psychologists will now come rushing at me in great force and with great conviction. Their hordes of followers will surround me to stone me or pray for me or debate with me or psycho-analyze me or numb me to death with their rhetoric. Gurus who have their own answers held with unwavering mass appeal and who write bestselling books which I can condense into the content of just one sentence, or to concepts already known to the Greeks, will nonetheless continue to prosper. But it seems to me that I am a skeptic, doubtful of any final answers, incredulous at those who espouse final answers, and certainly therefore not of disposition to propose any. But, like anyone, I do have hopes and wishes.

At the quantum mechanics level, I have found myself within the hidden variables persuasion. Like Einstein, and Josef Jauch, I would prefer to search for causative mechanisms underlying our current probabilistic understandings of quantum mechanics. This search would not contradict the probabilistic models which we have and which valuably describe quantum mechanics. Maybe my preference toward a deeper search is just a manifestation of my natural curiosity. However, from the classic book of G. Thomson, *The Atom*, I am compelled to quote: "Chance is in truth only another name for ignorance of the causes." I would even be happy to find deeper underlying cause which itself perches upon some finer underlying sea of chance. But evidently, if our lives are to have meaning, I would prefer that meaning to be related to something better than tiny random motions, whether or not they be set up by God.

How about at the macroscopic level? If, indeed, the three elements of *cause, chance,* and *choice* are at all times operating within the universe, do divine mechanisms operate them, and may we claim to be able to see or know those mechanisms? Consider my departure as a child from Iowa to Colorado. Against my beloved aunt, I do not believe that it was God's specific choice that I be uprooted from a relatively dull but secure life as a child in Iowa to a less secure but more stimulating life in Colorado. Macroscopic causes and chance human trajectory crossings would seem the operative elements. But I cannot be sure. Aunt Leona could be right. Or a more general *karma* of the Hindus and Buddhists may have been the guiding hand.

As it currently stands, a theoretical physicist could say that God has implemented *cause* via Newtonian classical mechanics, *chance* via Planckian quantum mechanics, and *choice* in the free-will decisions we may make in our personal lives. It is helpful to use the quantum mechanical notion of mixed states to think of all three elements of *cause, chance,* and *choice* as always present, in principle, at every instant of our life's trajectory. Our lives proceed on causative schedules. Eating causes energy causes activity. These actions are like classical mechanics. They essentially always work to sustain our existence. *Chance*, like quantum mechanics, is always present although usually not obviously so. *Chance* on the macroscopic level provides most of the richness of our lives, a richness of human trajectories which are massively interwoven. Those trajectories take place upon a large event space of unimaginable extent. *Choice* can enter by advantageously recognizing an event or flow of events as those of good luck and accepting them and acting upon them. Granted, one cannot necessarily make such choices when one is asleep, although we cannot rule out choices made when dreaming.

* * *

Awareness and openness should be maintained as often as possible in order to perceive and appreciate and act upon this great richness of our lives. I happen to believe, as a matter of faith, that God's event space is so large as to be beyond our imagination of it. So one meaning of it all is to be as open as possible to sensing the great beauty of this cosmic experiment into which we have been placed, as both actors and spectators. Another meaning is that we are certainly free to try to understand it, whatever it is. And thirdly, we may act upon it.

The ever thoughtful Greeks already dealt with much of this. Of special interest is Plato's dialogue of the astronomer Timaeus with Socrates. As Timaeus warms up, he asks:

> "What is that which always is and has no becoming; and what is that which is always becoming and never is? That which is apprehended by intelligence and reason is always in the same state; but that which is conceived by opinion with the help of sensation and without reason, is always in a process of becoming and perishing and never really is. Now everything that becomes or is created must of necessity be created by some cause, for without a cause nothing can be created..... But the father and maker of all this universe is past finding out; and even if we found him, to tell of him to all men would be impossible. And there is still a question to be asked about him: Which of the patterns had the artificer in view when he made the world – the pattern of the unchangeable, or of that which is created?"

Thus, all of *cause, change*, and *choice* are ever present, and we may be forgiven if we cannot answer the second question of this chapter. Later Timaeus argues that the soul of the universe was put in place before the universe itself, but then he imposes upon his own reasoning the caveat:

> "but this is a random manner of speaking which we have, because somehow we ourselves too are very much under the dominion of chance."

Throughout his discourse Timaeus couches all of his conclusions as just probable ones. And when speaking of the primary constituents of the physical universe, he identifies the inherent appearances of complexity and our need to deal with it:

> "Hence when they are mingled with themselves and with one another there is an endless variety of them, which those who would arrive at the probable truth of nature ought duly to consider."

Then there was Heraclitus, who said:

> "Aeon is a child at play, playing draughts; the kingship is a child's."

This may be paraphrased after interpretation to:

> "Time is a child playing dice; the Kingdom belongs to the child."

Notes

I never published my 1972 rudimentary Trajectory Analysis notes because they were far too subjective and incomplete. I was not alone. As recounted by Manfred Kochen, editor of *The Small World*, Ablex Publishing, New Jersey, 1989, Kochen credits Ithiel de Sola Pool for the initiation of a mathematical study of the small world problem in 1957. Kochen and de Sola Pool then circulated their ideas for two decades before publishing them in 1978.

A useful recent article on the topic was written by Mark Newman: *The physics of networks*, Physics Today, November 2008, 33–38. I find it interesting that in his discussion of assortive mixing, he notes that the probability of a connection between two individuals has been observed to depend on their ages, incomes, races, and the languages they speak. These are attributes like those I had written down in 1972.

The mathematics of Google search engine may be found in the book *Google's Page Rank and Beyond*, Amy Langville and Carl Meyer, Princeton University Press, 2006.

Our papers on context are *Contextual quick-learning and generalization by humans and machines*, Network: Comput. Neural Syst. 9 (1998), 85–106, and *Ergodic Learning Algorithms*, in Unconventional Models of Computation (eds. C. Calude, J. Casti, M. Dinneen), Springer (1998), 228–242. The paper on irreversibility was, I. Antoniou and K. Gustafson, *From probabilistic descriptions to deterministic dynamics*, Physica A **197** (1993), 153–166.

The Einstein quotations may be found in *The Quotable Einstein*, Alice Calaprice, Princeton Press, 1996.

Timaeus may be found in *The Dialogues of Plato*, Volume 7 of the *Great Books*, University of Chicago and Encyclopedia Britannica, Inc., 1952. Heraclitus is presented in *Probabilistic causality and Irreversibility: Heraclitus and Prigogine*, Theodoros Christidis, in Between Choice and Chance (eds. H. Atmanspacher, R. Bishop), Imprint Academic, Thorverton, U.K., 2002, 165–187.

14. Realities

...and time is running out...

The realities of our lives always take over. They are inescapable as long as we are here, breathing, existing, thinking. One may wax philosophic about their meaning or delve scientific into their origin. But such abstractions are really a leisure activity. The realities of our lives actually determine, describe, and thereby define much of the meaning of our lives.

Personally, I am fascinated at being still in the game. But I am also aware that time is running out. My age is certainly the overriding reality of my life these days.

But wait: there is a caveat here, that of a powerful and more universal reality. Your life, no matter your age, could be over before mine, even within a year, even tomorrow. So the blunt truth is that time is running out for everyone from the instant they are born. And individually, there are never any guarantees; we cannot know our future.

One of the sophomoric mottos in our high school climbing group was: A long life may not be good enough, but a good life is long enough.

A less sophomoric view of the fundamental uncertainties that rule our fates comes from The Rubaiyat:

> Some for the Glories of this World
> And some sigh for the Prophets Paradise to come
> Ah, take the Cash, and let the Credit go
> Nor heed the rumble of a distant Drum.

Here the uncertainty referenced is the afterlife. The corollary message is that the one reality that matters is the quality of my life, your life.

On the other hand, if I was counting on the Prophets Paradise to come, then that belief would surely dominate my life, and my age would be relatively inconsequential. If you have read this far, you can undoubtedly surmise from this book that this is not the case. But I am always searching for answers...am still searching. That has been a continual reality of my life.

What are the realities of my life these days? Have I learned anything about myself? As I cast my gaze back and draw it forward, I think of "these days" as beginning about five years ago, on May 6, 2005. My daughter Amy and her young family had arrived to visit from California, and she in her confident manner said, "Dad, organize a 70th birthday party for yourself." I replied, "Fine. You have given me one day's notice." Undaunted, Amy countered, "Just call a few friends who are around." So I called a few friends and the next day we had a delightful little fest in my small back yard and then as night fell, we moved inside. As all sipped wine or beer or whatever their choice, my daughter proposed a small toast to me. "With no warning I asked Dad to have this small party. With his usual aplomb, he pulled it off. Thank you all so much for coming." For those of you who may have perspicacious daughters, it is gratifying to receive some such true appreciation in your later years.

I am an improviser. I am not an initiator. That reality comes out throughout the accounts of this book, and it has surprised me. A high school climbing buddy would say, "Hey, Gus, you want to try...?" and I would hop up and say, "Sure. Let's go!" I maneuvered my way in college from Chemical Engineering to eventually Mathematics (as John Nash also did). Women generally initiated my relationships with them. So did my other friends. I have shunned managerial roles, but when they have arrived upon me, I have adapted and dispatched them with competence...but perhaps more tellingly, with efficiency. A close friend once observed to me (astutely): "Karl, you are not an operator. You do not aggress. And it is almost impossible for you to tell a lie."

In truth, I found it difficult to write this tale of my journey through this life. Throughout the writing of this book I have met with unexpected upwellings of emotion. Life is an emotional business, or at least it should be. What meaning has a mechanical robot? We all try to maintain a somewhat even keel, but it is the ups and downs of our lives that bring out its meaning. I know this.

Yet the reality is that I am not introspective by nature. And I do not like to dwell upon my feelings. I suppose I am a typical male in that regard. I am just naturally extrospective: I am more interested in the world around me than in self-examination. Still, writing one's autobiography does require a measure of the latter.

* * *

Although we never know our destiny, nonetheless we seem to have enough free will to at least influence events – provided that fate allows. My dream had always been to spend my later years in Boulder, and that has happened. Remarkably, when I park my car at the university and walk west toward the Mathematics building, my steps trace exactly the same path I walked 60 years earlier as a boy of 14. I walk not to University Hill Junior High School (the building is still there, now an elementary school) but to the role of professor at the university. Instead of knowing only Iowa and Colorado, I have seen the world and played science at the highest levels. I am a happy grandfather of five grandchildren between the ages of three and eight.

Apparently I am a survivor. The reality of my upper middle class upbringing in Iowa gave me an intellectual advantage. The hardships of my rough lower middle class life and climbing in Boulder induced a certain stoic toughness. Evidently I am a risk-taker. But some instinct always seems to pull me back from the edge just in time.

These days my mind is sharp and my intellectual curiosity is as alive as it was in that child in Iowa. My knees are gone but not yet replaced, so I am still able to climb a mountain and then struggle down more slowly than going up. The students in my classes keep me on my toes. Computing systems I now view as a frustrating but necessary evil, even though I was one of their original creators. I continue to out-publish my current colleagues; for example, 60 published papers in the last 10 years, 160 in the last 20 years. Almost every colleague I knew in the world community of mathematicians and scientists has retired or died. Although I decline many invitations, I still give invited international plenary lectures from time to time. In the last five years I have given lectures at international conferences in New Zealand, Vietnam, Belgium, Canada, Sweden, France, Slovakia, and China.

The most recent invitation (as of this writing) took me to the IWMS2010 conference at the Shanghai Finance University in China. That engagement spawned two others at Shanghai University and Fudan University. As I had already traveled much of mainland China in my six lectures there sixteen years earlier, I decided this time to just stay, hang out if you will, in Shanghai for two weeks. The conference and the three lectures lined up nicely and I found myself with three completely free days at the end of my sojourn there. So those days I explored the city alone via the fine Metro system. The overpowering force of the human numbers, their industriousness and the massive construction going on in the city...the happy couples dancing to soft music piped out to the street on a Saturday night...China is well on its way and is a world force to be reckoned with.

One does not choose one's invitations as a plenary speaker. Eventually they will fade away as a new generation wants its own. But the reality is that even though I'm not in any of the "we invite you-you invite us" mafias, still I have value. I like that: both the fact of having genuine intrinsic value, and my not needing extrinsic mafia connections. It is gratifying.

* * *

To further guide this snapshot of my current life and interests, I invoke here a visual model which I created some years ago. I call it "The Three Dimensions of Life," and here's how it works: You take the usual three orthogonal coordinate axes of the Euclidean three-dimensional world in which we live. Place yourself at the origin. Label the axis that points in front of you: Professional. Label the axis pointing to your left: Parent. Label the axis pointing above you: Private. The scope of the professional axis should be clear. The parent axis is to include also general family concerns. And the private axis will include your passions, hobbies, things you like to do as an individual. The merit of my model comes through more vividly when you now try to plot your life upon it. Stated simply, how is your energy spread over the three coordinates? Further, the model becomes dynamic as you trace how you must constantly shift your energies to meet your responsibilities and those interests which make life worth living.

In mathematical terms, your life is a moving vector within the positive orthant of this three dimensional space. To get a feel for this picture, go stand in the corner of a room, facing into the corner. Take a few deep breaths and focus in your mind how you are this day, multi-tasking your professional, your parental, and your private life.

Then slowly turn around and, with the right wall the professional, the left wall the parental, and the vertical corner across from you the private, direct your gaze according to the relative amounts of each of your three P's currently taking your energies. That is your meaning for today.

This way of looking at things, which I sometimes call "The Threefold Way," has been with me for some time. It is clearly evident in the Chapter 9, "Wives, Lovers, Friends." You must continually and dynamically adjust your efforts and even your heart and mind within the space of those coordinates. If you look at my first book on partial differential equations (1980), you see that the entire book is built completely on a threefold structure of trinities. And in Chapter 13, "The Improbabilities" here it seemed quite natural to me to place God within such an admittedly provisional threefold way of cause, chance, and choice. The unavoidable conclusion is that this threefold geometrical paradigm has become an archetype within my mind.

It is consistent with my 1972 Trajectory Analysis, which I described in the previous chapter. Just place your personal three dimensional life orthant on top of the currents of the world that are governing your life and let go and let it flow. Your orthant will have of course some predictability for the immediate future, just due to your normal daily schedules and habits. But all those possible future interactions with the orthants of millions of others provide the fascination of our existence.

What have been those big currents upon which much of my life flowed and therefore from which it took much of its meaning? I can identify three: the Satellite Revolution, the Cold War and in particular my role in Espionage, and of course that ever-present undercurrent of Uncertainty.

* * *

I think you will agree that the Professional "P" dimension of my life remains alive and well. I feel fortunate to be in a profession where that is possible. I recently turned down a generous buy-out offer that the university quietly floated, hoping to tempt some 50 tenured faculty to retire early. The university is under extreme budgetary pressures due to the current Depression. Rumor has it that only about 20 (out of a faculty approaching 1,000) took the offer.

Why give up your life? We who are world-class research professors are like world-class marathon runners. Like marathon runners, we train all the time. All the time…even in unpaid summers we are training all the time. But we even improve with age!

* * *

We turn now to the Parental "P" dimension of my current life. I had always looked forward to the day when "all the kids are gone" and I would be free of the responsibility for their care. However, after graduating high school, my son Garth felt comfortable continuing to live at home as he experimented with jobs such as fry cooking, sound engineer, and auto mechanic. He stayed 10 more years until he married. We were and are very comfortable together. So I had a fine roommate for a while.

I greatly enjoy being Grandpa Karl now. There are currently five grandchildren, three in California, and two nearby. Garth's two daughters Ashley and Elizabeth, 8 and 5 at this writing, are growing up fast. My daughter brings the other grandkids, Francesca, 6, and three-year-old twins Julian and Clarissa, here often for visits. Soon they will have less interest in Grandpa Karl as they develop their own identities and meanings. So I lap it up while it lasts.

I have been in the same house for more than 40 years. I managed to keep it through thick and thin. My view of the mountains is one of the most spectacular in Boulder. The house was architecturally designed to have an interior harmony with the outside world. We all feel a natural comfort whenever we assemble here. It is still home base. I could not have a better abode. I love this place.

I also have a small mountain cabin, built on an old mining claim beside a lake at nearly 10,000 feet. It has been a great benefit for the children, and especially for Patrick and Garth and their friends. One of Garth's friends had overdosed in high school, and his mother asked that his ashes be placed up there where he had had so many great times.

This cabin and the little cul-de-sac valley where it sits have a special quality. The location's isolation provides habitat for bear and moose, which we often see, and mountain lion, whose tracks we see from time to time. Somehow, as you enter this valley, you feel a quality I could characterize as – forgiveness. I go up there and find that my soul is soothed. I often take work but usually it just sits on the cabin table as I wander around the valley and hills in meditative wonder.

A continuing tradition now for over 30 years is to take everyone up there, especially now the grandkids, in mid-December to cut our Christmas trees. The cabin is in about half a mile and is snowbound six months of the year. Our tree-cutting expeditions may include sunshine in brittle cold or blizzard, but it is always memorable! Afterward there is the getting warm food at a good Mexican restaurant close to where my son and his family live.

Then there is the Christmas Eve cheese fondue at my house, now in its 35th consecutive year. Some years are lean with only a few of us here. Sometimes we also invite close friends. It is always an epic as Grandpa Karl needs a lot of help getting it right each year! For example, Christopher may take over the cheese grating, Garth the bread cutting, Stacy or Amy the fondue stirring, someone else the table setting. And the big question each year is always, how much wine should we add to the five pounds of Gruyere, Appenzeller, and (important) Vacheron cheese? What great times. Life is, as Hemmingway said, a moving feast.

Once a parent, always a parent. We are currently in a major economic Depression. The rich bankers overdid their gambling and their losses poured down in torrents onto the middle class and the poor. My son was unemployed for a year. His wife started a small childcare service in their home. Now he has a job better than the one he lost. But in the interim I was happy to supply some money to take up the slack in their very real needs. I really enjoy that and also take them all to dinner and elsewhere, anytime they want. I am a bit of a socialist at heart and I am happy to share my relative prosperity. Some friends have told me that I am overly generous. If so, so be it. I believe the Republican credo of the self-made-man is a myth.

A better description would be self-righteous forgetter of all who helped along the way. Generally I see the Achilles heel of Republicans to be this over-self-righteousness; and the Achilles heel of the Democrats to be a selective hypocrisy.

It seems I have absolutely no complaints about the "P" parental and family dimension of my life, both past and present.

Now let us look at the third "P" dimension of my current life, the private dimension. My health is good except for steadily progressing arthritis in my knees. The left one is worse. I destroyed it skiing in longthongs at age 20 before safety bindings were known. Then I destroyed it again playing basketball at age 40. But half of it is still there. A crisis occurred in 2001. I hiked too far, too high, with my friends Kathy and Brenda on a beautiful sunny blue-sky day in October just before the big snows came. We were in about seven miles when my bad knee gave out, and soon after on our way out, the good knee too. I came out the last three miles on poles constructed from tree branches. Brenda wanted to call an evacuation helicopter but I would not have it. By the time we were out, all six major joints – knees, hips and shoulders – had essentially frozen up. My doctor told me to get a new knee and sent me to the orthopedic surgeon. The X-rays showed bone on bone. But I hesitated. "Let me get the shoulders unfrozen first," I said. That took about six months. During that time the hips stopped hurting and the good knee came back. Four months of acupuncture may have helped. I went back to the orthopedic surgeon and told him I had decided "to not get a new knee, at least not just now." He was extremely honest and said, "Well, you have made the right decision. You would not be able to do the same things." This confirmed my view of my aging mountaineering friends who swore by their new hips and new knees...but whom I had observed were not doing things exactly as before.

I will capitulate eventually, I suppose. But not just now. I have replaced jogging with yoga the last five years and I enjoy that activity very much – even though about 20 percent of the postures are beyond my knee joints.

I lead a simple life. Breakfast is cold cereal, orange juice, and coffee. Lunch is often a brown bag in my office to allow a lunch hour of swimming and yoga. I prepare decent and even tasty dinners, and have no habit of desserts. If I am feeling a little low for some reason, I will cook up a batch of spaghetti with sausage meat and a good tomato sauce, along with a glass of a red table wine, and I am invariably instantly happy again. I have never been a TV person although I will watch it at dinnertime for the news and weather. I like adventure books, and during the financial meltdown I bought and read every good book on the subject. I detest mysteries and consider them a total waste of time. I usually have light FM rock or popular music playing softly in the background when I am home. I do not yet own a DVD player nor a cell phone.

I admit to some loneliness since my cat "Monsieur" died at age 18 two years ago. We happily spoke French together and were "toujours d'accord." But I have resisted bringing in some other, possibly French-speaking, creature....

One answer to the spiritual needs of my life has been my love of the outdoors. And I cannot live without mountains. Their stark beauty brings physical reality very close to my soul. If you insist that I choose a church, then it will be the halls of the mountain gods. Perhaps the manner in which my spirit is refreshed there resembles that of the Navajo Indians upon their homeland.

We really don't change much. What we have found interesting and agreeable, remains interesting and agreeable. But there are some advantages to age. As one is freed up from the worries that tend to accumulate around our three P's, for example the financial pressures that each of them tend to bring on...well...why not be happy? As the great French playwright Moliere wrote: "It is a stupidity second to none, to busy oneself with the correction of the world." Because I have a natural inclination to jump in to try to solve problems, be they professional, parental, or private, to counter those tendencies I have taken his words to heart. How did I do this? Speaking very candidly, about ten years ago I just stopped trying to "save" the Mathematics Department from what I perceived to be its follies. On the parental front, I keep my mouth shut and do not give advice but always try to be supportive in whatever way the kids want. In my private life I decided one day to just "go off the sex standard" and take what comes, and not need what does not come. By the way, the metaphor here for the younger reader is "going off the gold standard" – as the United States and other major currencies did about forty years ago. Let the currencies achieve their natural economic market values.

You cannot easily make such liberating decisions in midlife unless you have shirked or otherwise avoided the usual responsibilities that naturally accrue to your three P's in that period of life. But I can. Sometimes I joke that I am a liberated professor, a liberated grandparent, and a liberated lover.

Am I optimistic? I can do no better than quote my friend Ilya Prigogine in an interview he gave the Brussels newspaper Metro in 2003, a few months before his death. I translate: "Optimism, what does that really mean? The universe remains mysterious and very complex. We are part of a world which astonishes us. No one among us has chosen to be alive."

I am alive. Somehow I was given good health and a quick and imaginative mind. Isn't that enough?

Psychologically, I have been on my own ever since I was 13 and my parents' lives and thoughts went elsewhere. To survive and then enjoy life I developed a certain independence of spirit and mind. Philosophy has always interested me but it must be on my terms, and productive. Sometimes I think of myself as a warrior, like a samurai. Perhaps we can take that warrior posture into all of the three "P" dimensions. Professionally I have been a warrior and not a supplicator. Parentally I have been a brave single parent who also took responsibility to save his mother, his aunt, a stepson, and several friends. Privately I have battled mountains and have known beautiful women. Now I continue on. I continue the Tao, wherever it takes me.

> I do not know whether I was then a man,
> Dreaming I was a butterfly,
> Or whether I am now a butterfly,
> Dreaming I am a man.
>
> –Chuang-Tsu

Notes

The full-page interview which Professor Prigogine gave the Brussels newspaper Metro appeared June 2, 2003.

The Rubaiyat of Omar Khayyam was translated into English by Edward Fitzgerald in 1889 (first edition). My edition of The Rubaiyat was published by The Peter Pauper Press, Mount Vernon, New York.

15. The Crossing of Heaven

His skis were running well now. When he started out, the trail had been icy, with just a sprinkling of new snow; but as he headed higher into the mountains, that new snow of last night was now an inch deep. Why had he not skied during the last five years?

His mind wandered back to his last girlfriend. They had adventured for fifteen years; but then she had taken up telemark skiing...and he could not. His knees had absorbed a lifetime's beating from mountain climbing, then mountain running, and then even jogging had to be suspended. The orthopedic surgeon wanted to replace his bad knee but he had said, no thanks, not right now. So she, who had always valued her independence, "just telemarked away," as he liked to say. There had been no goodbyes...they had just drifted apart.

Then, his work, the creating of new mathematics, had filled his time. Not retired, he also enjoyed the teaching and social interactions at the university. So let's call it just inertia. The inertia of life was enough. So there had been no skiing.

Recently he had begun to write his autobiography. No one knew that in 1960 he had written the software for the world's first spy satellite. No one knew that his mathematical algorithms had allowed fast tracking of the Russian submarines in the Cuban Missile crisis of 1962, when the world was brought so close to nuclear war. He wanted to tell those stories. So he had hired an editor to help him. And she, in her soft, intelligent way, suggested that he not work so hard all the time, and "don't forget to sneak off skiing." That was last night. This morning he had grabbed his twenty-year-old cross-country skis, poles, and boots, and headed up to the high country.

He knew exactly where he would go: to one of the most strikingly beautiful places. To get there, he would have to ski in for about three miles in deep mountain woods at about 10,000 feet. He was in about one mile now. Surprisingly, he had the trail to himself. No tracks preceded his as he glided and climbed upward.

Then he noticed the coyote tracks which had joined the trail. Great, he thought, this fellow is no dummy...why not walk along the packed track? They continued together for about another mile. Then the slope steepened. Still the coyote of that morning stayed on the track. The snow was about two inches. deep now and perfect as his skis cut their path through it. With a little heavy breathing, suddenly he emerged from the trees, and he was there!

The Crossing of Heaven. Great wind-sculpted white snowbanks mixed with small green wind-stunted alpine conifers, under the deepest of blue skies. He looked west up the valley to Mount Toll, 13,000-feet high, the first mountain he had climbed, at age 14. He looked up the other valley at Navajo Peak, 13,400 feet, the second mountain he had climbed, the weekend after his first ascent. He was all alone. Unable to restrain himself, he shouted into the eternity before him: "Heaven! Heaven! Heaven!"

There was no answer. He marveled at the stillness. It was unusual. The Crossing of Heaven was a giant windtrough where the hurricane-force western winds roared over and down from the divide, and up over this small ridge, and then hurtled eastward toward the Great Plains. Many times he had had to battle his way across to the shelter of the trees on the other side. Before him was this barren ground into which the coyote's tracks vanished. No snow could survive there. He kept to the snowdrifts and found himself on the other side and in about 15 minutes he skied down a small trail to the closed and shuttered mountain club cabin nestled in the trees.

Only an hour and 15 minutes. Why, that was the same time as five years ago! Of course, there had been no talking, no stopping, this day. One must eat. One must always eat in this arctic world. He found a small stump-seat in the sun and pulled out the emergency food he had hurriedly stuffed into his pack. One-year-old cheese crackers and raisin cookies and a little water and then he rose to put his skis on to begin the descent. A 15 minute sojourn lunch and to stay longer would mean to get cold and stiff.

Quickly he was back at the Crossing of Heaven. Looking at the great barren trough, he wondered where the coyote was going. Still no wind. For five minutes he could not tear himself from this incredible place, the great divide peaks rising like a giant wall covered with fresh snow, which would be torn from them and funneled through this crossing and then deposited eastward.

Finally he turned his back to the divide and entered the trees and started smoothly down the wonderful snow on the trail winding through the forest. He accelerated as the slope steepened and his mind went back to the Crossing and then to how well he felt after not skiing for five years and then...he was down in the snow! He laughed at himself and untangled his skis but found that he could not get up. How does one get up? It occurred to him that he was totally alone. Swinging his skis below him, he managed to push up onto one of them, then onto the other. Skiing more carefully, he made his way back to the trailhead. The Crossing of Heaven would always be there when he needed it.

GPSR Compliance
The European Union's (EU) General Product Safety Regulation (GPSR) is a set of rules that requires consumer products to be safe and our obligations to ensure this.

If you have any concerns about our products, you can contact us on

ProductSafety@springernature.com

In case Publisher is established outside the EU, the EU authorized representative is:

Springer Nature Customer Service Center GmbH
Europaplatz 3
69115 Heidelberg, Germany

www.ingramcontent.com/pod-product-compliance
Lightning Source LLC
LaVergne TN
LVHW050013270326
834688LV00068B/35